AF768872

Microscopic Analysis of Agricultural Products

Fourth Edition

Editors: James Makowski
Neil Vary
Marjorie McCutcheon
Pascal Veys

AOCS PRESS

AOCS Mission Statement
To be a global forum to promote the exchange of ideas, information, and experience, to enhance personal excellence, and to provide high standards of quality among those with a professional interest in the science and technology of fats, oils, surfactants, and related materials.

Library of Congress Cataloging-in-Publication Data
Microscopic analysis of agricultural products / editors: James V. Makowski . . . [et al.]. -- 4th ed.
p. cm.
Previous eds. published as: Manual of microscopic analysis of feedstuffs.
Third ed. published 1992.
Includes bibliographical references and index.
ISBN 978-1-893997-71-4 (alk. paper)
1. Feeds--Analysis. 2. Feeds--Microscopy. I. Makowski, James V. II.
Title: Manual of microscopic analysis of feedstuffs.
SF97.M53 2010
636.08'550287--dc22

2010036795

Printed in the United States of America
15 14 13 12 5 4 3 2

The paper used in this book is acid-free, and falls within the guidelines established to ensure permanence and durability.

Contents

Chapter 1. Methods of Agricultural Microscopy 1

Fundamentals of Agricultural Microscopy

This chapter, intended for beginning agricultural microscopists, describes the principles of ingredient identification, equipment used, sample preparation techniques, observation methods, and related topics.

Flotation

One of the most widely used methodologies for sample separation is flotation. In this procedure, feed samples are separated into density-dependent fractions by adding aliquots of two solvents of different specific gravities. The procedure results in samples that can be arranged in homogenous, density-dependent fractions.

Chapter 2. Feed Ingredients of Plant Origin 9

This chapter, as with subsequent sections on animal and marine ingredients, utilizes definitions recognized by the Association of American Feed Control Officials (AAFCO). This does not preclude the use of other ingredient definitions. This section discusses key stereomicroscopic and cellular or subcellular identification features.

Chapter 3. Feed Ingredients of Animal Origin 43

This chapter examines the key stereoscopic and cellular and subcellular identification features of feed ingredients derived from animals. Identification features of ingredients such as dairy products, meat meals (bovine, ovine, porcine, equine), and poultry products are discussed.

Chapter 4. Feed Ingredients of Marine Origin 55

As agriculture increases, so do the types of feed ingredients found in aquaculture diets. This chapter discusses major ingredients found in aquaculture diets, such as fish meals, shrimp meals, crab, meals, etc. As in previous sections, both macroscopic and cellular aspects are examined.

Chapter 5. Detecting Animal Products in Feeds and Feed Ingredients 65

This chapter deals with the detection and identification of various animal protein products. As a result of the emergence of bovine spongiform encephalopathy (BSE), or "mad cow disease," and the link between this disease and feeding practices, government bodies throughout the world (e.g., the U.S. Food and Drug Administration) have put restrictions on the use of animal protein products in animal feeds. A separate section dealing with the detection and identification of specific markers, such as hair and bone, which can be used to identify animal protein products, is included.

Chapter 6. Fertilizer Microscopy 99

This chapter examines various components/ingredients of fertilizers and how they can be identified using both visual and microchemical markers.

Chapter 7. Weed Seeds of Agricultural Importance 115

The identification of weed seeds is important to agricultural microscopists. This chapter provides agricultural microscopists with photographs and descriptions of some of the more common weed seeds.

Chapter 8. Minerals of Agricultural Importance 135

The mineral chapter provides microscopists with photographs and descriptions of some of the more common minerals found in animal feeds. This section is recommended as a supplement to microchemical spot testing for identification of specific minerals.

Chapter 9. Microchemical Spot Tests 161

This chapter details the methods of spot testing for a variety of minerals, vitamins, and drugs common to feed rations. The section includes principles of spot testing, reagents, steps for each test, results to expect, and other pertinent information.

Appendix A: Glossary 171

This section contains a glossary that defines the special nomenclature of agricultural microscopy drawn from the various scientific disciplines. Suggested reference sources and related information are also listed.

Appendix B: Mesh Size Conversion Table 175

Bibliography 177

Contributors 181

Index 185

Preface to the Fourth Edition

The American Association of Feed Microscopists (AAFM) published the first edition of *The Manual of Microscopic Analysis of Feedstuffs* in 1955 (revised 1960), the second edition in 1978 and the third edition in 1992. These manuals have served veterinarians, toxicologists, scientists, and technicians in agriculture for more than 50 years.

Recognizing the tremendous increase in the area of agricultural microscopy, the Agricultural Microscopy Division of AOCS, formerly AAFM, saw a need for a new, much expanded edition. We have deliberately changed the name of the manual to *The Manual of Microscopic Analysis of Agricultural Products, Fourth Edition* to reflect our desire to provide a more inclusive approach to the microscopy of agricultural products. The area of animal protein products and their detection in light of recent outbreaks of Bovine Spongiform Encephalopathy (BSE), or "Mad Cow Disease," as well as other Transmissible Spongiform Encephalopathies (TSE's), requires a much more detailed methodology for detecting the presence of animal protein products. Returning to our roots, so to speak, we have included coverage of fertilizers and expanded the number of unusual feed ingredients.

Agricultural microscopy is both an art and applied science that draws from many academic disciplines. Plant and animal anatomy, feed technology, and analytical chemistry are just a few that have contributed. Thus this manual is designed as a study guide for the beginning microscopist and as a reference for the more experienced. It is not a complete treatise. It is a compilation of practical information that required many hours of labor by individual microscopists and others working toward common goals of agricultural product identification.

So many people have contributed to past editions that credits can no longer be equitably given. Because efforts are all voluntary, each edition represents the collective knowledge of editors and committees. Participants in this current manual are richer for their many hours spent, both as contributors to those who will follow and in honor of those who built the basic framework for agricultural microscopy. We submit our collective effort in this manual as a guide for all students and practitioners of agricultural microscopy. Please learn and enjoy the ideas and information presented here, then share your experiences and observations with other microscopists.

Acknowledgments

The editors would also like to especially thank those microscopists and past editors who came before us and provided the groundwork in the previous editions, which we built upon for this manual edition:

Bob Bax (deceased), St. Louis, MO, USA.
Egerton Whittle (deceased), Athens, GA, USA.
Gary Lewis, Watkinsville, GA, USA.
Lewis Barefield, Little Rock, AR, USA.
Lynn S. Bates, Manhattan, KS, USA.
Randy Sample, Ft. Wayne, IN, USA.
Scott Webster, Quincy, IL, USA.
Sue Jaconis, St. Louis, MO, USA.
Wynne Landgraf, Ames, IA, USA.

The editors would like to acknowledge the following people, companies, and institutions, without whom this manual could not have been produced:

Canadian Food Inspection Agency, Ottawa Laboratory (Carling), Ottawa, Ontario, Canada.
Community Reference Laboratory for Animal Proteins in Feedingstuffs, CRA-W, Belgium.
Evan Balaskas, Canadian Food Inspection Agency, Ottawa Laboratory (Carling), Ottawa, Ontario, Canada, photography.
Geneviève Frick, Agroscope Liebefeld-Posieux Research Station, Switzerland, weed seed consultation.
Jose Hernandez@ USDA-NRCS PLANTS Database, photography.
Kenneth Howe, Holt, Michigan, USA, line drawings.
Mark W. Skinner@ USDA-NRCS PLANTS Database, USA, photography.
Messiah College, Grantham, PA, USA.
P.S. Wright (deceased), Florida Department of Agriculture and Consumer Services, Tallahassee, Florida, USA, Fertilizer Microscopy Chapter.
Steve Hurst@ USDA-NRCS PLANTS Database, USA, photography.
USDA, NRCS, National Plant Data Center, Baton Rouge, LA 70874-4490, USA.
Walloon Agricultural Research Centre – Valorisation of Agricultural Products Department, Gembloux, Belgium.
West Virginia Department of Agriculture, Regulatory and Environmental Affairs Division, Charleston, WV, USA.
Windsor Laboratories, Mechanicsburg, PA, USA.

Illustrations and Tables

The illustrations in this fourth edition of the Manual, including photographs, photomicrographs, line drawings, maps and tables originated from several sources, who kindly provided permission to AOCS Press to reuse their original material. Each original source is credited in the caption of the illustration; and AOCS Press would like to further acknowledge these generous contributions from the following sources, which are listed in alphabetical order:

Evan Balaskas, Canadian Food Inspection Agency, Ottawa Laboratory (Carling), Ottawa, Ontario, Canada, photography.
Canadian Food Inspection Agency (CFIA), Ottawa Laboratory (Carling), Ottawa, Ontario, Canada.
Community Reference Laboratory for Animal Proteins in Feedingstuffs, CRA-W, Gembloux, Belgium.
Geneviève Frick, Agroscope Liebefeld-Posieux Research Station, Switzerland, weed seed consultation.
Jose Hernandez@ USDA-NRCS PLANTS Database, photography.
Kenneth Howe, Holt, Michigan, USA, line drawings.
Steve Hurst@ USDA-NRCS PLANTS Database, USA, photography.
Mark W. Skinner@ USDA-NRCS PLANTS Database, USA, photography.

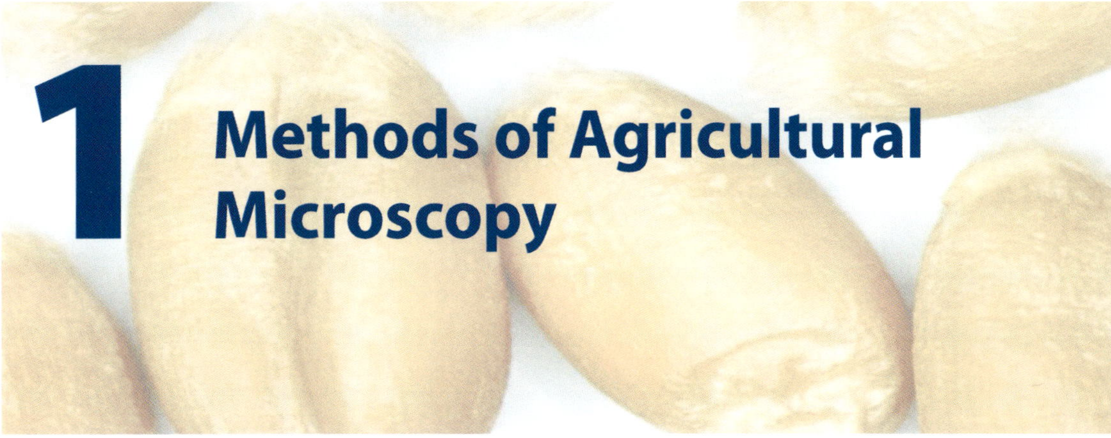

1 Methods of Agricultural Microscopy

Principles of Ingredient Identification

Agricultural microscopy is the art and science of identifying ingredients in agricultural and agrifood products using stereo and compound microscopy. An ideal complement to chemical and chromatographic analyses, agricultural microscopy is a rapid, relatively inexpensive tool for determining the quality of raw ingredients and finished agricultural and agrifood products. The goal of agricultural microscopy is to provide manufacturers and consumers a technique for determining which agricultural ingredients and minerals are present in a given product, and that the feed, fertilizer, or food is nutritious and safe for animal or plant maintenance and production, or human consumption.

Microscopic identification of agricultural products is based on comparisons of unknown materials to known standards. The essential skills of the agricultural microscopist center on the identification of the ingredients found in these products, both singularly and in complex mixtures. These include the various forms of each ingredient (high quality, low quality, underprocessed or overprocessed, etc.) along with contaminants and adulterants. This task is compounded by the tremendous heterogeneity of these products.

Agricultural microscopy methodology for the identification of agricultural products can be divided into two major types—macroscopic and cellular/subcellular. Macroscopic identification, the method used by a majority of agricultural microscopists, employs a stereoscope, or "dissecting" microscope, generally ranging from 5× to 50× magnification, to identify and evaluate ingredients and foreign material via the physical characteristics of an ingredient. This method requires little sample preparation, and is dependent on the microscopist's skill in identifying ingredients by their shape, color, particle size, softness, hardness, texture, luster, odor, etc.

The second method, the identification of ingredients by their cellular or histological structures using a compound microscope, relies on the microscopist's ability to identify tissues, cellular structures, and arrangement of plant and animal materials. Although requiring more technical skill, identification of cellular and subcellular features provides more accurate results since the identity of cellular and subcellular features is not lost, even when an ingredient is reduced to powdered form. The ingredients being examined are typically immersed in a mounting medium and then examined at a magnification ranging from 40× to 400×, depending on the configuration of the microscope. A magnification of 100× is usually adequate for most uses.

Either of these two methods may be used independently, but more accurate, reliable results will be obtained from the use of both methods.

Commercial agricultural and agrifood products are comprised of plant, animal, and marine products. Most ingredients are processed (e.g., ground, crimped, rolled, dried, cooked, etc.). Therefore, the microscopist should become familiar with various plant, animal, and marine tissues as well as the effects of processing on these tissues.

Reference Ingredient Collection

Since microscopic identification of agriculture products is based on comparisons of unknown materials to known standards, every microscopy lab should have a collection of ingredients, adulterants, and contaminants, relevant to the commodities to be examined, for comparison to the constituent(s) found during microscopic analysis. This type of reference library can be built by requesting samples from feed/fertilizer/food manufacturers, by-product producers, government seed and feed control laboratories, feed/fertilizer/food brokers, agricultural and agrifood colleges, and other microscopists. Participants in AOCS-sponsored agricultural microscopy short courses receive an initial set of samples on which to build a working inventory of ingredients found in these commodities.

Reference samples should be studied by the microscopist until he/she is familiar with each of them. Again, recognizing ingredients in these commodities by comparing them to a known reference is the basis of agricultural microscopy. Significant time spent collecting, maintaining, and observing an orderly system of reference standards will contribute significantly to agricultural microscopy expertise.

Instruments and Equipment

The type of equipment used in an agricultural microscopy laboratory depends on the amount and degree of anticipated use, the personal preferences of the user, and budget. No hard and fast rule will be given, but essential items, as well as other items that have proven useful, are described.

Microscopes are a major and long-term investment. Because of the hours spent using microscopes and the need for clear definition of ingredients examined, the quality of the microscope directly affects the accuracy of the analyses obtained. Both a stereoscope and a compound microscope are needed for best results.

A stereoscope with wide-field eyepieces, objectives, and a magnification range of 10–40× is recommended. Parallel optics are preferable to convergent systems in order to reduce eyestrain. A zoom objective is preferable to separate objectives for rapid work. The single most important factor affecting how well one sees objects through the stereoscope is the light source. Poor or low light sources will not allow for observation of fine surface detail and will quickly lead to eye fatigue. Therefore, choosing a proper light source is critical. Although many stereoscopes have built-in light sources, it may be best to choose one without, allowing the use of external, adjustable lighting. If budget is not a factor, consider halogen light "pipes." This arrangement allows for white, bright light, and the ability to bend the flexible "pipes" to achieve the best lighting or shading. If under budgetary constraints, a fixed halogen light, even a simple desk lamp, may be adequate. Halogen is the light source of choice, since it produces the whitest, brightest light.

A compound microscope for fine-particle analysis and cellular and subcellular confirmation is essential for the agricultural microscopy laboratory. The compound microscope should have a binocular head to allow viewing with both eyes and 10× plan achromatic (flat field) oculars. Coupled with plan achromatic 4×, 10×, 40×, and 100× objectives, the microscope will be able to produce total magnifications of 40×–1,000×. Plan achromatic oculars and objectives are recommended, as they allow for the entire field of view to be in focus at one time, preventing continuous focusing and refocusing of objects at the extremities of the field. A mechanical stage and adjustable condenser and diaphragm are important for proper contrast adjustments. Polarizing elements and a rotating stage are highly recommended to allow the microscopist to examine birefringent features of agricultural and agrifood ingredients. Although the compound microscope represents a significant investment, it will last many years and provide the microscopist with the opportunity to observe and confirm particles of the smallest size. In many instances, compound microscopy may be the only method to accurately identify agricultural and agrifood ingredients.

In addition to a stereoscope and compound microscope, the minimal basic equipment for the agricultural microscopist should consist of the following:

1. A mechanical or electronic balance, capable of 0.01 g accuracy will allow for precise weighing of samples as they are prepared for analysis, and for precise weighing of flotation/separation fractions for estimating the percentages of ingredients found in the samples.
2. An ocular micrometer with a stage micrometer to calibrate it, to allow the microscopist to measure the size of particles under the compound microscope.
3. A syringe (infants' rubber), brush, lens paper, cotton swabs, and alcohol for keeping instruments clean and dust free.
4. Sieves of small diameter with screens approximately 10, 20, 30, and 40 mesh, plus a cover and pan. These may be used individually to separate larger from smaller particles, or arranged in order of largest to smallest mesh to produce a size-dependent array of particles.
5. Evaporating dishes, a method for removing heavy solvent vapors, and simple laboratory stirrers for flotation/separation methodologies (see Sample Preparation by Flotation).
6. Other handheld instruments, such as stainless steel forceps, dissecting needles, and a stainless steel micro spatula to aid in sample handling and manipulation of individual particles while using the stereoscope.
7. Some microscopists keep a small brush, scalpel (or single edged razor blade), and spot plates at their workstation. Small containers, such as petri dishes, small beakers, disposable aluminum pans, and plastic or paper cups are also helpful.
8. Mortar and pestle to break apart conglomerates of ingredients without destroying individual ingredient particles.

An agricultural microscopy laboratory can be set up in a small office or alongside an existing quality assurance laboratory. The equipment is relatively inexpensive for the information it can generate within a short period of time. Because of the vast range of ingredients present in agricultural products, learning this material from textbooks and manuals is difficult. The Agricultural Microscopy Division of the American Oil Chemists' Society (AOCS) offers basic, advanced, and specialty short courses to help train agricultural microscopists. Learn more at www.aocs.org.

Preparing the Sample for Inspection

For dependable results, a representative portion of the sample is essential. Thoroughly mix the sample to provide an accurate representation of the ingredients present. If the sample is pelleted, it should be reduced by crushing either in a mortar or a flour mill before mixing. Take care to crush and not tear the sample particles in order to allow their structures to be identified under the stereoscope. Hand mixing of a sample is accomplished by spreading the material on a large sheet of paper and rolling it crosswise (toward opposite corners), supplemented by stirring with a large spatula. The sample can be reduced to a workable size by quartering and taking diametrically opposed quarters until the required amount is obtained. A riffle sampler can be used in place of the steps described. Sample amounts will vary with the fineness of grind; more of a coarse feed is required than a finer meal.

Sample Separation by Flotation

Flotation of samples to separate inorganic from organic constituents of ingredients may be applied before or after sample size reduction. (Dense organic particles can separate into the heavy fraction and vice versa—e.g., bone fragments may be found in the inorganic, or heavy fraction.) Flotation separates the sample into a series of density-dependent fractions (see Flotation Section). Flotation also enhances the visibility of particles by removing fats and oils from the sample. Sieving after flotation yields cleaner separation of ingredients by particle size.

Sample Separation by Sieving

Sieving the sample can be very helpful to the agricultural microscopist. Sample particle size and the expertise of the microscopist dictate the number and mesh size of sieves used. Sieving through a set of 10-, 20-, and 30- or 40-mesh sieves separates the sample into four fractions. The separation of coarse particles from fine particles augments the examination of ingredients that would be difficult to recognize if the finer particles were not removed.

Inspection with the Stereomicroscope

The flotation separation procedure will result in several density-dependent fractions of known weight. Sieving a sample will produce several size-dependent fractions. In both procedures, the resulting fractions should then be transferred to an observation board or paper and examined under the stereoscope. Regardless of the procedure used, transferring particles to an examination board or paper should be done whereby larger particles are transferred first, followed by medium-sized particles and then fine particles. This can be accomplished by tipping the fraction toward the examination board and then gently tapping the fraction from underneath with a small brush or your finger. After some practice, the microscopist can begin to arrange each fraction in such a way that the largest particles will fall onto the board first, followed by medium, and finally fine particles. Try to arrange the particles so that they are in a single layer, which will aid in preventing smaller particles from being hidden by larger ones.

To observe a fraction, place the examination board or paper under the stereomicroscope and, under 10× or 20× magnification, begin at one edge of the fraction and work across. As you examine the particles under the stereomicroscope, use a needle or fine forceps to manipulate individual particles for closer inspection, and to test individual particles for hardness, texture, structure, and other identifying characteristics.

If the material being examined is a mixed feed or fertilizer, and a list of the ingredients is available, each ingredient should be checked off as it is identified. Unlisted ingredients should be recorded as they are encountered. If, after examining and listing all ingredients thought to be present, no unlisted ingredients have been identified, it is good practice to re-examine the sample, paying particular attention to the fines. If no list of ingredients is available, record each ingredient as it is identified, using official or generally accepted terms. Adulterants are often finely ground to escape detection. If the product is coarsely ground or poorly mixed, a second portion should be prepared and examined. The absence of a listed ingredient is confirmed when both the second examination and the fines observed at 100× magnification with a compound microscope are negative for that ingredient.

The procedures described previously can be used to identify the most common mineral ingredients, particularly those added to feeds and fertilizers in appreciable amounts and not ground to powder form, as with some salt, calcium, and phosphorous supplements. The presence of trace minerals, added in proportionately smaller quantities, and usually of smaller particle size, can be confirmed by chemical spot tests. These tests are often conducted under the stereoscope because some of the reactions may be difficult to detect with the unaided eye. The microscopist will find mineral identification and spot tests easier to perform if extraneous material is removed by flotation. See Chapter 9, Microchemical Spot Tests, for additional information.

Inspection with the Compound Microscope

In the hands of a competent microscopist, the previously described procedures with the stereomicroscope will prove adequate to identify most of the ingredients, contaminants, and adulterants in formula feeds and fertilizers. When ingredients are present in very small quantities or have been ground so finely that they are difficult to find or identify with the stereoscope, use a compound microscope.

Scanning a sample at 100× magnification is fast and satisfactory for most purposes. Higher magnifications may be used if needed.

Only a few milligrams of sample dispersed in a drop of mounting medium are needed for examination with the compound microscope. Too much sample limits light transmission, interferes with clear observations, and makes ingredient identifications difficult. Also avoid using too much mounting medium. Several mounting media may be used: water, glycerol, mineral oil, or a 60% (w/v) aqueous chloral hydrate solution. The latter is perhaps the simplest media to use. A 1:1:1 (v/w/v) mixture of distilled water, chloral hydrate, and glycerol or 100 g crystalline phenol in 20 mL of glycerol is also good mounting media. Placing ingredients in mounting media and heating will help to displace trapped air and allow some ingredients of finer cellular structure to be more readily observable. Always compare the unknown ingredient to known reference standards before drawing conclusions.

A mechanical stage is desirable to allow rapid and precise scanning of ingredients under the compound microscope. Systematically examine any slide or preparation by beginning at the upper left or right margin and moving down the width of the slide, one field of view at a time. Move the slide one field of view across and then proceed up the width of the next area, again one field of view at a time. Continue in such a pattern until the entire slide has been searched.

Bone and mineral constituents can be examined after flotation and drying of the heavy fraction. The dried heavy fraction can be transferred to a petri dish, or placed onto a piece of black or cobalt blue glass, and manipulated in such a way as to separate the particles by size. The finest particles can be directly observed in mounting medium. Larger particles of bone or other opaque particles should be broken, crushed, or cut, if possible, and smaller particles examined.

Quantitative Estimation of Ingredients by Flotation/Separation

Much of the routine work of the agricultural microscopist is qualitative. The presence or absence of ingredients or contaminants with estimates of ingredient quality is often the most critical aspect of a rapid quality assurance program. Increasingly, however, clients are requesting the percentage of a specific ingredient, or a complete percentage breakdown of a mixed product.

The quantity of sample to use for quantitative estimations varies according to the nature of the product under study. For example, for most feeds in the meal, pelleted or crumbled form, 2–10 g is ample. For feeds consisting of pellets, meal, and whole grain, at least 50 g should be used. This type of feed, often referred to as a textured feed, is first screened or handpicked into its various components, which are then treated separately.

Pelleted or otherwise agglomerated products must be ground by hand in a mortar or by using a ripple faced flour/grain grinder. The object is to crush the materials into individual ingredients without tearing or destroying fine features. Avoid use of a hammer mill or bladed grinder, such as a coffee mill.

Follow these steps to separate mixed ingredients into density-dependent fractions:

1. Weigh a representative 2 g sample (grind with a mortar and pestle or pass through a flour hand grinder if necessary) into an evaporating dish.
2. Add enough chloroform ($CHCl_3$) to cover the sample (6–12 mL) and stir well. After the mineral and/or bone settles, transfer the floating material (by pouring and with the aid of a spatula), plus the $CHCl_3$ into another evaporating dish. Allow the mineral/bone fraction to dry. This is the “heavy” fraction.
3. Add a sufficient amount (2–4 mL) of petroleum ether (or hexane) to the $CHCl_3$ to the transferred material. Stir well, and then transfer the floating material onto a filter paper. This is the lightest fraction.
4. Add 2-mL aliquots of $CHCl_3$ until 2–3 additional fractions are obtained. The resulting fractions will be produced in lighter to heavier order.
5. Dry all fractions. If quantification is desired, weigh each fraction using an electronic or mechanical balance to the nearest 0.01 g. Once a fraction is weighed, transfer

it onto a dark colored examination board, ensuring a single layer of material. Examine each fraction under the stereoscope.

NOTE: use chloroform and petroleum ether with caution. Perform separations under a fume hood or with a laminar flow hood, wear gloves, and avoid breathing vapors or contact with skin.

The end result of these procedures will be 4–6 density-dependent fractions, which are then weighed using an analytical balance. Transfer each fraction to an examination board so as to display the fractions in a single layer of particles, decreasing in size from largest to smallest (fines) particles (Tables 1.A and 1.B).

The first step in quantitative estimation is identification of all ingredients present. Therefore, prolonged study of individual ingredients is necessary before any sort of estimation is possible. After all ingredients are listed, estimate them one at a time, studying all fractions and the whole sample, remembering to take into account each fraction weight. For example, if a particular fraction comprises 28% of the total sample, and the microscopist estimates that 15% of that fraction is made

Table 1.A. Typical Complete Ration Separations[a]

Heavy	Medium	Light
calcium carbonate	alfalfa meal	corn gluten meal
phosphate	ground corn	corn gluten feed
salt	ground milo	citrus pulp
bone	ground barley	feather meal
sand	ground wheat	soybean meal
crystalline amino acids	wheat middlings	fine alfalfa
metals	ground oats	brewers grains
	ground beet pulp	cottonseed meal
	soybean hulls	urea
	sugar	cottonseed hulls
	ground rice	corn cob
	rice bran	rice hulls

[a]Some ingredients may be found in more than one fraction

Table 1.B. Meat Meal and Poultry Meal Separations[a]

Heavy	Medium	Light
bone	muscle tissue	muscle tissue
sand	hair	hair
glass	stomach contents	hide
calcium carbonate	intestinal contents	dried blood
metal shavings	fuzz (fluff)	raw feathers
egg shell	leather meal	hoof
oyster shell		horn, beak, claw

up of corn, then the amount of corn present in that fraction is 4.2% (28% × 15%). A quantitative estimate of the total corn present in the entire sample is achieved if the amount of corn present is estimated for each fraction and the percentage of corn present in each fraction totaled. Repeat this step for each ingredient listed until you have an estimate of all ingredients for that entire sample.

The agricultural microscopist should then re-evaluate the entire sample as a whole, taking into account the relative weight per volume of ingredients, the frequency of occurrence of each ingredient, and general knowledge of the characteristics of an ingredient (Table 1.C). Studying known mixtures of various ingredients can be very helpful. Initially, quantitative estimates can be reported as a spread. For example, the percentage of wheat middlings present in a feed sample may be reported as 9–12%. The correct level is assumed to be the median of the range, with possible variation to the high or low figure. The average of the high and low figures for all ingredients listed should be 100, +/−0.5%. Estimated protein levels can be determined by multiplying the expected protein of an ingredient by the percentage of that ingredient founding the sample. If chemical analysis of protein has been performed on the sample, the estimated protein can be compared to the tested protein for accuracy.

Much patience and study are required to do quantitative estimation in this way, as it is still primarily an art. However, surprisingly accurate results can be obtained, particularly after frequent practice.

Table 1.C. A Typical Finished Feed Formula

Ingredients	Approximate %	Calculated Protein
ground barley	2–3	0.33
ground yellow corn	60–65	5.20
ground oats	2	0.24
dehydrated alfalfa meal	1	0.10
corn gluten meal	1	0.60
wheat middlings	9–12	2.14
soybean meal	9–12	5.76
fish meal	2	1.20
poultry by-product meal	3	1.20
animal fat	2	0.00
NaCl	1	0.00
$CaCO_3$	1.5	0.00
PO_4	0.5	0.00
		16.77%

15.9% = Actual Protein Analysis
Range: 94–106
Average: 100

2 Feed Ingredients of Plant Origin

Introduction

For purposes of organization, this chapter is divided into three main categories: Cereal Grains, Oilseeds, and Miscellaneous Plant Materials. Each section begins with a brief description of the contents, followed by an alphabetical listing of those plant materials.

Cereal Grain Structure

Cereal grain species are grasses that produce single-seeded fruits. Botanically, this type of fruit is known as a caryopsis. The caryopsis is composed of the fruit case (pericarp) and the seed itself. These tend to remain tightly bonded together, even during most processing. The embryo and endosperm, which are the result of fertilization, are enclosed by seed coats and pericarp tissues. Shape, size, and mass are the most readily identifiable characteristics of the fruits of individual cereal species (Evars & Millar, 2002). However, considerable variation of each of these parameters can occur within species (Fig. 2.1).

Endosperm

The endosperm is made up of two tissues, the starchy endosperm and the aleurone. As the name implies, the starchy endosperm contains starch granules, which can be identified under the compound microscope and which allow one to distinguish between the various cereal grains.

Aleurone cells are high in protein and are composed of thick-walled, blocklike cells, and can aid in identification of cereal grains. Most of the aleurone layer is removed during milling.

Nucellus

Lying just exterior to the aleurone, the nucellar layer is the remaining tissue from which the embryo developed. While the nucellus may be quite prominent is some cereal grains (sorghum), it may be completely absent in others (oats).

Testa

The testa is the term used for the true seed coat, which lies just outside the nucellar layer. The thickness and number of testa layers varies within cereal grains. In mature sorghum, the testa is lacking. In wheat, triticale, and rye, the testa is composed of two layers; whereas

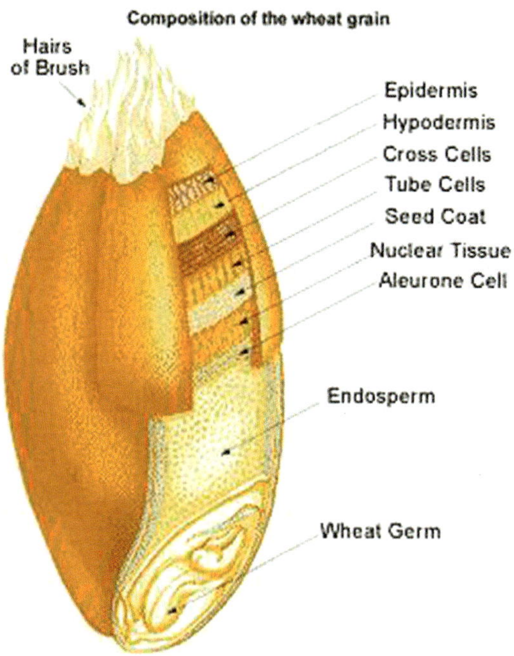

Fig. 2.1. Cereal grain (wheat).

in rice, oats, and barley, it is composed of only a single layer.

Pericarp

The outermost layer of the cereal grain is the pericarp, which is composed of dry, empty cells of varying shapes and arrangements. The innermost layer of the pericarp contains cells that have a tubelike or wormlike appearance, generally referred to as tube cells, and running parallel to the embryo's long axis.

Outside the layer of tube cells are cells found only in grasses and run perpendicular to the tube cells. These "cross" cells vary in shape and thickness and also serve as a means of distinguishing between cereal grains.

Barley Products—*Hordeum vulgare*

Barley is used widely in feeds as whole grain or in ground, crimped, or rolled form. Barley products are used in ruminant feeds more frequently than in poultry or swine feeds. Additionally barley by-products are covered in the Brewers Products section.

Definitions

Barley is the entire seed including palea and lemma which may be ground, crimped, or rolled.

Pearl barley by-product is the entire by-product resulting from the manufacture of pearl barley from clean barley.

Barley mill by-product is the entire residue from the milling of barley flour from clean barley and is composed of barley hulls and barley middlings.

Barley hulls consist of the outer covering of barley.

Macroscopic Features

Barley is a spindle-shaped, buff-colored, five-angled grain with a ventral groove. The kernel is hard with a light tan to grey outer lemma and palea (glumes). The hull is thin, lusterless, roughly textured, and adheres to the caryopsis until ground, when it breaks into characteristic triangular pieces. Barley may be awned or awnless (Fig. 2.2).

Fig. 2.2. Whole barley. © Her Majesty the Queen in Right of Canada (2010).

Stereoscopic Features

Ground barley is identified by the triangular particles of hulls with prominent ribs. Barley endosperm is floury and harder than oat endosperm. Broken awns also indicate the presence of ground barley (Fig. 2.3).

Crimped or rolled barley is identified by the ribs of the glumes adhering to the grain, the luster and texture of the hull, and the awn when present. Enough of the whole grain remains after crimping or rolling to distinguish barley from oats, wheat, rye, or triticale. Other barley by-products can be identified by using the above features.

Fig. 2.3. Barley hulls. © Her Majesty the Queen in Right of Canada (2010).

Histological Features

The cells of the outer layer of the glumes are elongated with wavy walls, twin cells, and round cells. One of the twin cells is oval and the other crescent-shaped (Fig. 2.4).

Epicarpal cells are elongated, polygonal, and indistinctly beaded except at the apex where they are smooth and form rosettes around the hairs. These hairs have lumens narrower than their walls and range up to 300 μm long (Fig. 2.5). The cross cells are a double layer of transversely elongated, unbeaded cells found over the greater part of the kernel. Intercellular spaces occur sparingly between the end walls, especially in the outer layer. The tube cells are detached and form the endocarp.

Aleurone cells and starch cells comprise the majority of the endosperm. The aleurone cell layer differs from most cereals by having

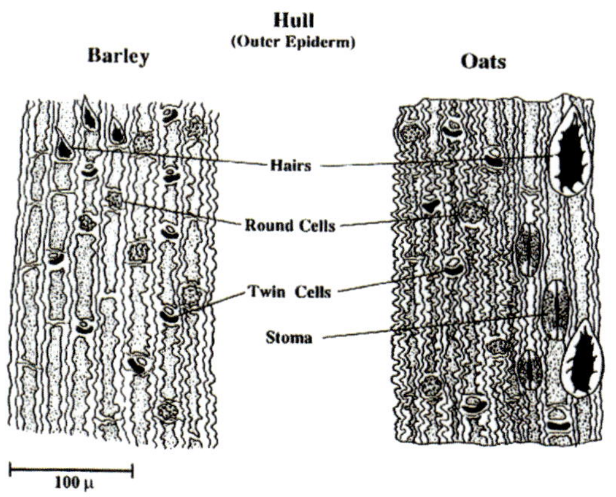

Fig. 2.4. Hull (outer epiderm).

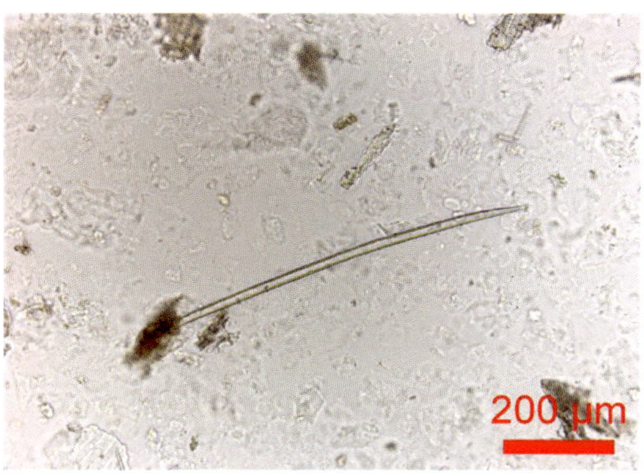

Fig. 2.5. Barley trichome. © Her Majesty the Queen in Right of Canada (2010).

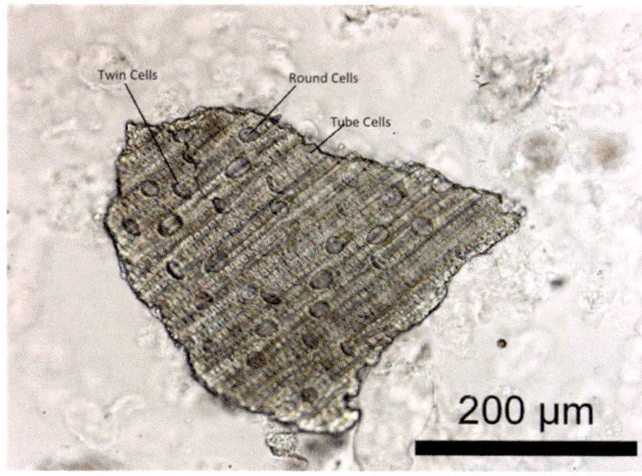

Fig. 2.6. Cross, twin, and round cells of barley. © Her Majesty the Queen in Right of Canada (2010).

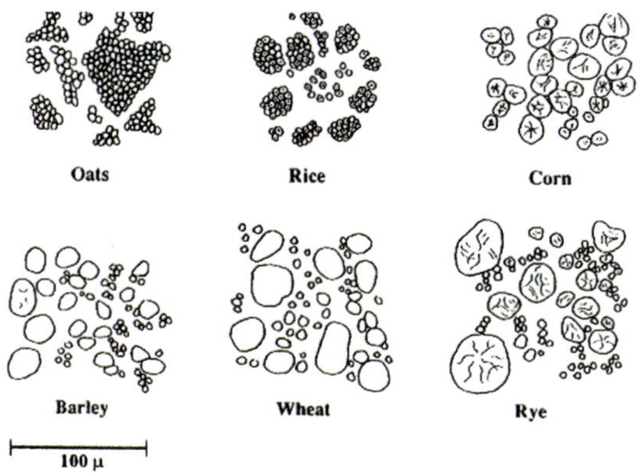

Fig. 2.7. Starch grains.

two to four cells instead of a single layer. The starch grains are similar to wheat starch, are lens-shaped, and seldom exceed 30 μm in diameter (Figs. 2.6 and 2.7).

Corn (Maize) Products—*Zea mays*

Corn is fed in whole and milled forms and as by-products to all types of livestock. Brewers and Distillers by-products from corn are covered under those sections.

Definitions

Whole corn is the entire ear with kernels on the cob and covered with the husk.

Whole shelled corn is the complete kernel removed from the cob without grinding.

Corn bran is the outer covering of the corn kernel with little or none of the starchy part of the germ.

Ground corn, cracked corn, corn meal, or corn chop is the entire corn kernel ground or chopped. Foreign material must not exceed the AAFCO definition.

Corn germ meal (dry milled) is ground corn germ with other parts of the corn kernel from which part of the oil has been removed. It is obtained in the dry milling manufacture of corn meal and other corn products.

Corn germ meal (wet milled) is ground corn germ from which most of the solubles have been removed by steeping and most of the oil removed by hydraulic expeller or solvent extraction. It is obtained in the wet milling manufacture of corn starch, corn syrup, and other corn products.

Corn gluten feed is that part of commercial shelled corn that remains after the extraction of the larger portion of the starch, gluten, and germ by the processes used in the wet milling manufacture of corn starch or syrup. It may contain corn germ meal and/or fermented corn extractives.

Corn gluten meal is that part of commercial shelled corn that remains after extraction of the larger part of the starch and germ, and separation of the bran by the processes used in the wet milling manufacture of corn starch and syrup, or by enzymatic treatment of the endosperm. It can contain corn germ meal and/or fermented corn extractives.

Heat processed corn is the product resulting from heating whole corn under pressure (steam processing) or with dry heat (dry roasted) without removing any of the component parts. It may be ground, pelleted, flaked, or crimped.

Ground ear corn, corn, or cob meal is the product obtained by grinding or chopping the entire ear of corn, without husks, with no greater portion of cob than occurs in the ear corn in its natural state.

Ground ear corn with husks is the product obtained by grinding or chopping the entire ear of corn with its husk, with no greater portion of the cob and husks than occurs in the ear corn in its natural state.

Fig. 2.8. Hominy feed. © Her Majesty the Queen in Right of Canada (2010).

Flaked corn is the product obtained by running cracked corn, which has been aspirated and properly tempered, over smooth flaking rolls after which it is dried and cooled. If toasted after drying it is called ***toasted corn flakes.***

Ground corn cob is the product resulting from grinding the entire corn cob.

Hominy feed is a mixture of corn bran, corn germ, and part of the starchy portion of either white or yellow corn kernels, or a mixture thereof, as produced in the manufacture of pearl hominy, hominy grits, or table meal. If prefixed with the words "white" or "yellow," the product must correspond thereto (Fig. 2.8).

Corn grits or hominy grits are the fine to medium hard particles obtained from ground corn. They contain very little bran or corn germ.

Kibbled corn is obtained by cooking cracked corn under steam pressure and extruding it from an expeller or other mechanical pressure device.

Corn feed meal or maize corn fines is the fine siftings from ground or cracked corn, with or without aspirated products added back.

Macroscopic Features

The corn kernel is tooth-shaped and covered with a bran coat. The surface of the bran is cutinized (waxy appearing). When corn is ground, cracked, or crushed, its basic characteristics are not altered (Figs. 2.9 and 2.10).

Fig. 2.9. Whole yellow corn. © Her Majesty the Queen in Right of Canada (2010).

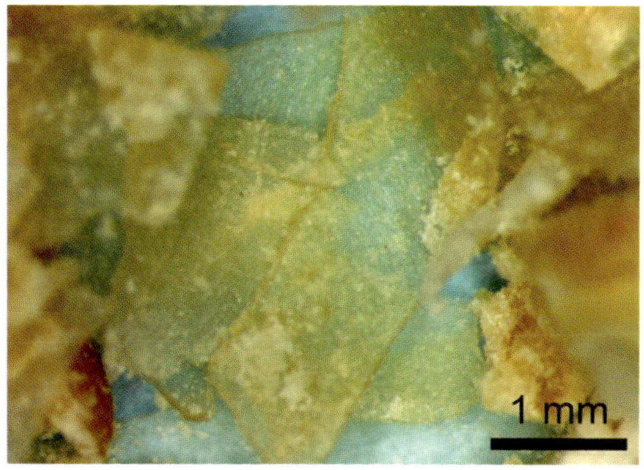

Fig. 2.11. Corn bran. © Her Majesty the Queen in Right of Canada (2010).

Fig. 2.10. Ground yellow corn. © Her Majesty the Queen in Right of Canada (2010).

Fig. 2.12. Corn glume or 'bees wings'. © Her Majesty the Queen in Right of Canada (2010).

Stereoscopic Features

Bran is glossy, thin, and semitransparent along with having faint linear ridges similar to those of fingernails. The horny or corneous endosperm is located primarily along the sides with a small amount along the back of the kernel. The white, floury endosperm portion surrounds the back side of the germ (Fig. 2.11).

The tip caps appear as funnel-shaped, woody pieces from the base of the kernel at the cob attachment site. Glumes (beeswings) are generally red or colorless and appear veined. They are especially abundant in corn screenings and corn cob fractions (Fig. 2.12).

Ground corn cob is identified by hard, woody lumps or irregular shaped pieces. Husks are characterized by strands of ropelike fibers with part of the connecting tissue attached.

Several corn by-products lack the characteristics of whole or ground corn. Corn gluten meal has no cellular structure unless contaminated with a small amount of bran or other corn particles. It is identified by its unique, slightly rough-surfaced (dull textured), golden spheres (Fig. 2.13).

Corn gluten feed is produced by blending wet gluten with bran and then drying. The product is characterized by corn bran that is darkened by the corn gluten on its surface. The corn gluten is not visible to the feed microscopist. Corn gluten meal may be present to balance the protein content (Fig. 2.14).

Fig. 2.13. Corn gluten meal. © Her Majesty the Queen in Right of Canada (2010).

Fig. 2.14. Corn gluten feed. © Her Majesty the Queen in Right of Canada (2010).

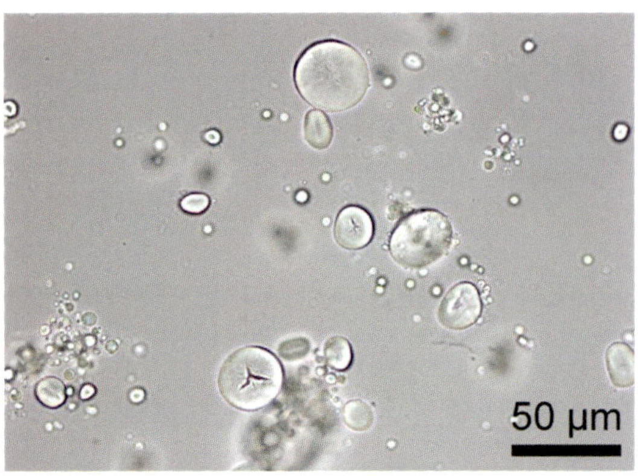

Fig. 2.15. Corn starch. © Her Majesty the Queen in Right of Canada (2010).

Histological Features

Cells in the starchy endosperm are elongated and generally arranged end-to-end. The size and shape of starch vary with the location in the endosperm. Large granules are found in the floury endosperm and smaller ones in the horny endosperm. Starch grains in the horny endosperm are polygonal; those from the floury endosperm are rounded. Both types range from 15 to 30 μm in diameter. Corn starch has a distinct central hilum, often with radiating clefts. This hilum distinguishes corn starch from wheat, oats, and barley; but grain sorghum starch is so similar that it is difficult to differentiate between them (Figs. 2.7 and 2.15).

Grain Sorghum Products—*Sorghum bicolor M.*

The grain sorghums (also known as milos) are widely used in animal rations. There are other sorghums grown for molasses and for forages.

Definitions

Ground grain sorghum is the entire product made by grinding the grains of grain sorghum.

Grain sorghum gluten feed is that part of the grain of grain sorghum that remains after the extraction of the larger part of the starch and germ, by the processes used in the wet milling manufacture of the starch or syrup.

Grain sorghum gluten meal is that part of the grain of grain sorghum that remains after the extraction of the larger part of the starch and germ and the separation of the bran by the processes used in the wet milling manufacture of the starch or syrup.

Grain sorghum germ meal or cake consists of the grain sorghum germ from which part of the oil has been pressed. It is obtained in the wet milling manufacture of starch or syrup.

Grain sorghum mill feed is a mixture of grain sorghum bran, grain sorghum germ, part of the starchy portion of grain sorghum kernels, or mixtures thereof, as produced in the manufacture of grain sorghum grits, refined meal, and flour.

Macroscopic Features

The presence of grain sorghum is usually indicated by whole or partial ground grains. The grains are small, oval to rounded, with a small point at the germ end. They can be brown, reddish-brown, white, yellow, or mottled. Another characteristic is the shiny black, yellow, brown, or purple empty glumes (Figs. 2.16 and 2.17).

Stereoscopic Features

In ground sorghum, the bran coat adheres closely to the grain after grinding. These ground particles are hard, lusterless, and off-white, with purple and pink streaks due to weathering. These small seeds break to give particles with one slightly curved surface.

Grain sorghum gluten feed is composed largely of bran and has a distinct odor similar to corn gluten feed. The bran is dark to reddish brown and usually curled at the margins (Fig. 2.18).

Grain sorghum gluten meal is the gluten portion separated from bran or other seed structure, although enough bran remains to use as a diagnostic feature. The meal separates into round or irregular shaped, hard, dark-brown spheres.

The shiny colored glumes may be ground and added to some sorghum products.

Fig. 2.16. Whole sorghum. © Her Majesty the Queen in Right of Canada (2010).

Histological Features

Starch granules in the pericarp are a unique characteristic of grain sorghum. The bran is more colorful than corn bran because the cells contain red, orange, pink, yellow, and reddish-brown pigments. The color is not uniform; dark streaks and patches give the bran a mottled appearance. Individual grain sorghum starch granules are indistinguishable from corn starch.

Grain sorghum gluten feed and meal bran particles contain air bubbles extending the length of the epidermal cells; this is not seen in untreated bran. The meal is opaque with translucent edges.

Fig. 2.17. Ground sorghum or milo. © Her Majesty the Queen in Right of Canada (2010).

Fig. 2.18. Sorghum or milo gluten feed. © Her Majesty the Queen in Right of Canada (2010).

Oat Products—*Avena sativa*

The cereal oat is widely used in both feeds and foods. In animal feeds, both oats and oat by-products are used extensively.

Definitions

Clipped oat by-product obtained in the manufacture of clipped oats. It may contain the light chaffy material broken from the end of the hulls, empty hulls, light, immature oats and dust. It must not contain any excessive amount of oat hulls.

Feeding oat meal is obtained in the manufacture of rolled oat grouts or rolled oats and consists of broken rolled oat groats, oat chips, and floury portions of the oat groats, with only such quantity of finely ground oat hulls as is unavoidable in the usual process of commercial milling.

Mixed feed oats consist of the entire product made by chopping or grinding a mixture of predominantly cultivated and wild oats.

Oats must contain a minimum of 80% sound cultivated oats and must not contain more than 10% wild oats.

Oat groats are cleaned oats with the hulls removed.

Oat hulls is the product consisting primary of the outer covering of the oat.

Oat mill by-product is the by-product obtained in the manufacture of the oat groats and consists of oat hulls and particles of the groat.

Macroscopic Features

Commercial oats are thin, elongated seeds, virtually round in cross section and tapering toward both ends. They are generally light yellow or buff, but may vary from almost white to weathered gray. Oats have a lustrous and thick hull, long hairs or trichomes, and soft white endosperm that distinguish them from barley (Fig. 2.19) .

Stereoscopic Features

When ground, oat hulls break into roughly rectangular fragments; barley hulls splinter into triangular fragments with jagged edges. Oat hulls are more flexible than barley hulls (Fig. 2.20).

Oat groats are identified by their softness (easily crushed) and the long silky hairs (trichomes) which give oats a downy appearance. The starchy endosperm has the consistency of a semi-dry paste that can be spread smoothly with spatula pressure. Other floury cereal endosperm will not spread smoothly (Fig. 2.21).

Oat trichomes are colorless, shiny, slightly curved, needle-like hairs scattered throughout the feed. They mat and form clusters when sieved (Fig. 2.22).

Fig. 2.19. Oats—whole grain. © Her Majesty the Queen in Right of Canada (2010).

Fig. 2.20. Oat hulls. © Her Majesty the Queen in Right of Canada (2010).

Fig. 2.21. Oat groats. © Her Majesty the Queen in Right of Canada (2010).

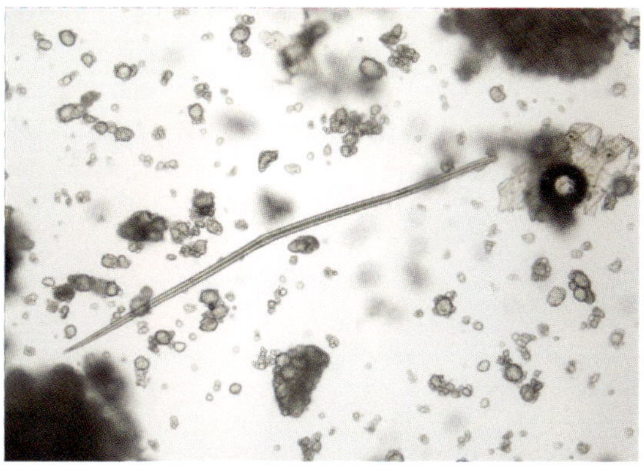

Fig. 2.22. Oat trichomes and starch. © Her Majesty the Queen in Right of Canada (2010).

Excessive numbers of trichomes indicate the presence of a lower quality material such as clipped oat by-product. An abundance of small, clean-cut, rectangular hull fragments indicates oat mill by-products. Common adulterants in oat products are other cereal grains, oilseeds, and weed seeds.

Histological Features

The outer epidermis of the oat hull, especially the outside of the central dorsal region, is composed of long cells with sharply defined zigzag walls. Barley outer epidermis cells are broader, less sharply defined, have wavy walls, and are thicker in proportion to the cell width. Both barley and oats have round cells and twin cells in the hull (Figs. 2.4 and 2.23).

Oat hairs have a lumen narrower than the walls, taper at both ends, range to 2 mm long, and have a bulbous base.

Individual starch granules are hard to differentiate from those of rice. Both are polygonal, small (2 to 20 µm), and have no distinct hilum. Both are synthesized as aggregates in the amyloplast: oat aggregates are found as whole ovoids and rice aggregates as broken fragments. These may appear clear to opaque, depending on particle size (Figs. 2.7, 2.24–2.26).

Fig. 2.23. Oat hull 400×. © Her Majesty the Queen in Right of Canada (2010).

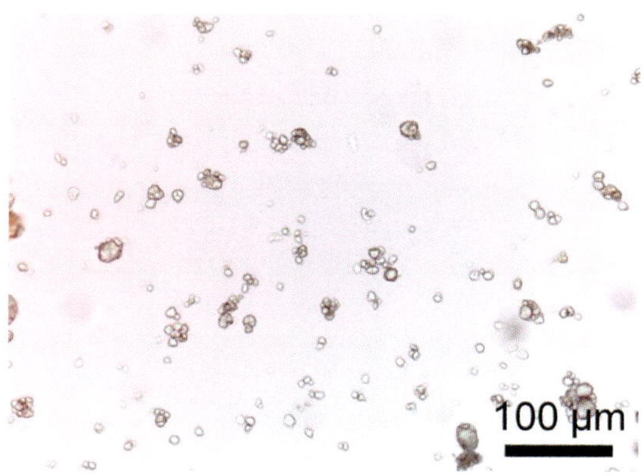

Fig. 2.24. Oat starch. © Her Majesty the Queen in Right of Canada (2010).

Fig. 2.25. Ground oats. © Her Majesty the Queen in Right of Canada (2010).

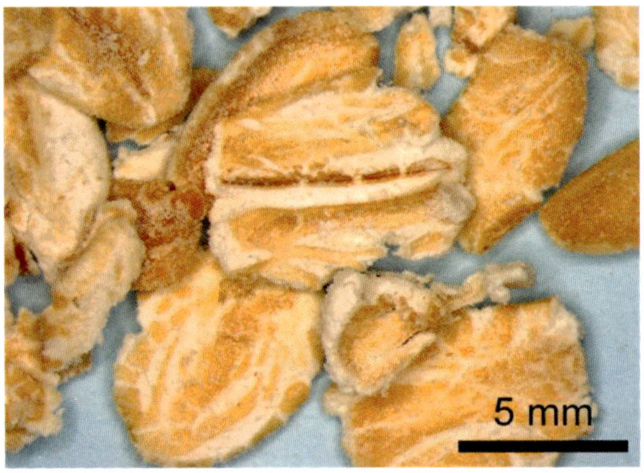

Fig. 2.26. Rolled oats. © Her Majesty the Queen in Right of Canada (2010).

Rice Products—*Oryza sativa*

Definitions

Ground rough rice or ground paddy rice is the entire product obtained in grinding the whole rice grain including the hulls.

Rice bran is the pericarp or bran layer and germ of the rice, with only such quantity of hull fragments, chipped, broken or brewers rice, and calcium carbonate as is unavoidable in the regular milling of edible rice.

Chipped rice, broken rice, or brewers rice is the small fragments of rice kernels that have been separated from the larger kernel of milled rice.

Rice hulls consists primarily of the outer covering of the rice.

Rice mill by-product is the total offal obtained in the milling of rice. It consists of rice hulls, rice bran, rice polishings, and broken rice grains.

Rice polishings is a by-product of rice obtained in the milling operation of brushing the grain to polish the kernel.

Ground brown rice is the entire product obtained in grinding the rice kernels after the hulls have been removed.

Macroscopic Features

The rice grain (caryopsis) is enclosed by the lemma and palea, which form the hull. The rice grain is smooth, long, and elliptical in cross section. The grain is shiny white, translucent to opaque, with two longitudinal parallel ridges on each flat surface. The embryo is approximately

Fig. 2.27. Whole rice. © Her Majesty the Queen in Right of Canada (2010).

Fig. 2.28. Rice hulls. © Her Majesty the Queen in Right of Canada (2010).

one-quarter of the fruit length and lies along the flat surface at the base of the grain. The hull is cream to light brown with distinct crosshatching on the outer surfaces. The hulls are rough and raspy to the touch (Fig. 2.27).

Stereoscopic Features

The distinct, shiny crosshatching of the hull is the chief characteristic at low power (Fig. 2.28). Most of the rice by-products, except rice polishings, will contain large enough fragments of hull and grain to be identified. Also, portions of the pedicel are frequently found.

Histological Features

The hull, bran, and starch grains are the predominant diagnostic features for rice by-products.

Under higher magnification the rice hull has longitudinal rows of large, elongated, thick, wavy-walled cells with many air spaces. These spaces appear black until mounting medium displaces the air.

Rice bran is semitransparent with characteristics of rice hull but much smaller and less distinct. The bran resembles very fine mesh (Fig. 2.29).

Rice polishings are predominantly rice starch with a trace of bran. Rice starch granules are very small, from 2 to 10 μm in diameter. These granules form aggregates with as many as 100 starch grains (Fig. 2.7).

Solvent-extracted rice bran has a more distinct cellular structure and no oil globules as compared to unextracted bran (Figs. 2.30–2.32).

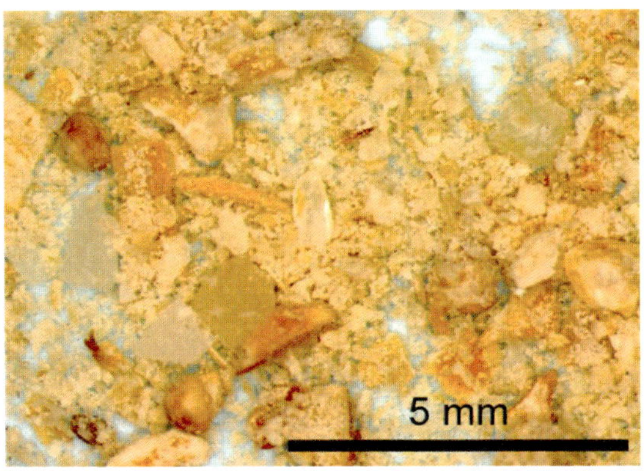

Fig. 2.29. Rice bran. © Her Majesty the Queen in Right of Canada (2010).

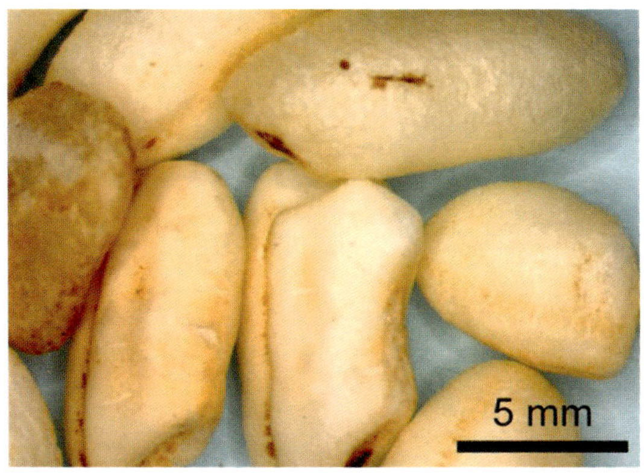

Fig. 2.31. Puffed rice. © Her Majesty the Queen in Right of Canada (2010).

Fig. 2.30. Polished rice. © Her Majesty the Queen in Right of Canada (2010).

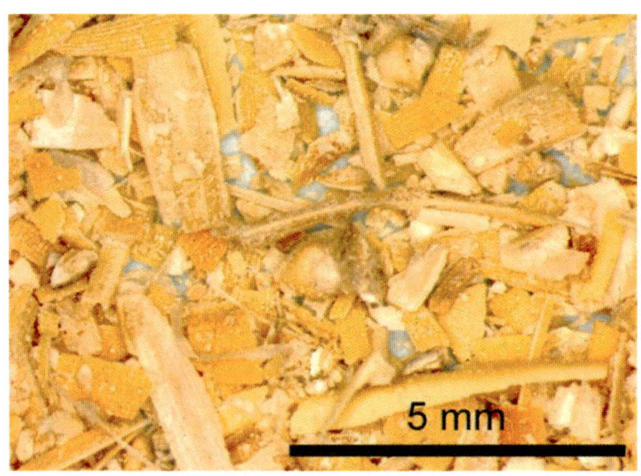

Fig. 2.32. Rice mill by-product. © Her Majesty the Queen in Right of Canada (2010).

Rye Products—*Secale cereale L.*

Rye is closely related to wheat but is used sparingly in feeds. By-products are more commonly found in mixed feeds.

Definitions

Rye mill run is the product obtained in the milling of rye flour from cleaned and scoured rye, consisting principally of the mill run of the outer covering of the rye kernel and the rye germ with small quantities of rye flour and aleurone.

Rye middlings consists of rye feed and rye red dog combined in the proportions obtained in the usual process of milling rye flour.

Rye bran is the coarse outer covering of the rye kernel as separated from the cleaned and scoured rye.

Fig. 2.33. Whole rye. © Her Majesty the Queen in Right of Canada (2010).

Macroscopic Features

The kernel is long and slender, pointed at the apex, and has a deep groove extending over the entire ventral side. The palea and lemma thresh free, leaving a dirty brown to greenish brown seed approximately 12 to 14 mm long (Figs. 2.33 and 2.34).

Stereoscopic Features

Rye is similar to wheat, but it is longer, more slender, and more tapered than wheat. It has greenish cast and the ventral groove is less prominent. The dorsal surface is more wrinkled than wheat.

Histological Features

In the pericarp (bran), epicarpal cells are longitudinally elongated. Their side walls are thinner than wheat with less beading and with solid end walls. This is the key feature to distinguishing wheat from rye.

Cross cells resemble those of the epicarp. These cells are indistinctly beaded only along the side walls; the end walls are thickened and rounded. Both the side and end walls of wheat cross cells are beaded (Fig. 2.35).

Rye starch is circular in surface view with radiating wrinkles and fissures. Rye starch granules often exceed 50 µm; few wheat grains reach this size (Figs. 2.7 and 2.36).

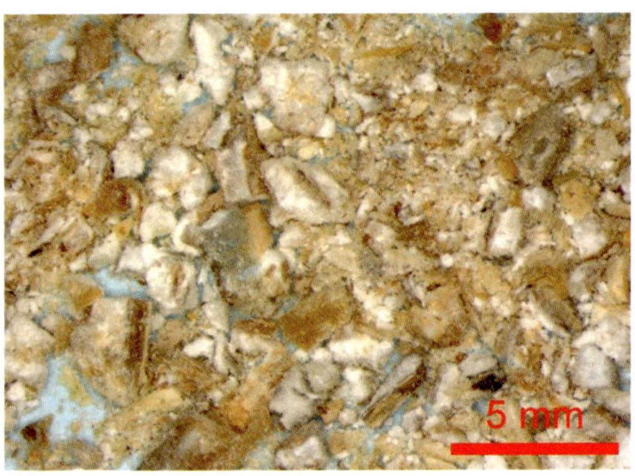

Fig. 2.34. Ground rye. © Her Majesty the Queen in Right of Canada (2010).

Fig. 2.35. Rye cross cells. © Her Majesty the Queen in Right of Canada (2010).

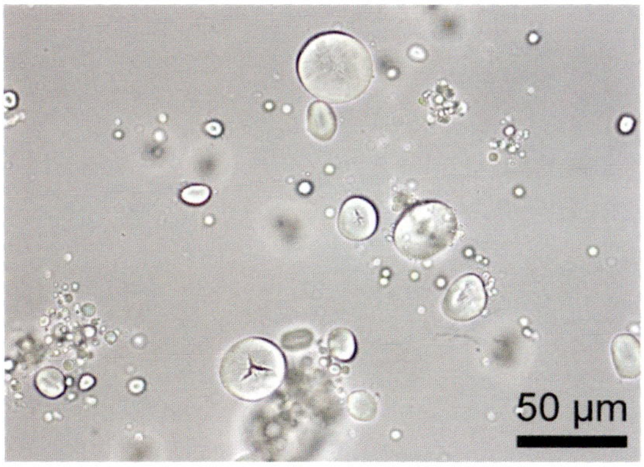

Fig. 2.36. Rye starch. © Her Majesty the Queen in Right of Canada (2010).

Wheat Products—*Triticum spp.*

Wheat milling, a mechanical separation of endosperm from embryo (germ) and bran, produces various by-products for the feed industry. These products are mixtures due to the milling process and subsequent blending for marketing.

Definitions

Wheat bran is the coarse outer covering of the wheat kernel separated from cleaned and scoured wheat in the usual process of commercial milling.

Wheat flour consists principally of wheat flour together with fine particles of wheat bran, wheat germ, and the offal from the tail of the mill.

Wheat germ meal consists chiefly of wheat germ together with some bran and middlings or shorts. If part of the oil is removed, it is referred to as ***defatted germ meal.***

Wheat mill run consists of coarse wheat bran, fine particles of wheat bran, wheat shorts, wheat germ, wheat flour, and the offal from the tail of the mill.

Wheat middlings consist of fine particles of wheat bran, wheat shorts, wheat germ, wheat flour, and some of the offal from the tail of the mill.

Wheat shorts consist of fine particles of wheat bran, wheat germ, wheat flour, and the offal from the tail of the mill.

Wheat red dog consists of the offal from the tailings of the mill together with some fine particles of wheat bran, wheat germ, and wheat flour.

Macroscopic Features

Whole wheat (Fig. 2.37) varies from light buff to brown. The kernel is blunt, deeply grooved, hairy at the tip, 6 to 8 mm long, and may be evenly rounded to irregularly heart-shaped in cross section. The proportion of horny to starchy endosperm varies according to variety; some are floury and others flinty.

The embryo is on the dorsal side of the grain and extends from base to one-third of the length of the kernel. Its position is indicated by shriveling of the surface.

The individual wheat products are off white to brown, bulky, flaky, floury, or mealy depending on the proportion of bran to starch. The more bran present or the larger its particles, the bulkier and flakier the product will appear. Conversely, less bran or finer bran particles will produce a more floury or mealy product. The proportion of bran to starch and the size of the bran particles depend upon the stage of milling from which it is derived.

Stereoscopic Features

Bran particles are found in most wheat products and are useful for identification. They

Fig. 2.37. Whole wheat. © Her Majesty the Queen in Right of Canada (2010).

range from white to reddish brown, are thin, finely wrinkled, and have tufts of trichomes called brush hairs (Fig. 2.38). White starch granules, appearing crystalline, adhere to the inner surfaces of the bran and provide a significant diagnostic feature (Fig. 2.39). Particles of whole ground wheat may be recognized by the adhering bran.

Histological Features

Bran and starch grains are the chief diagnostic features of wheat. The bran coat is composed of several thin layers; fragments with one or all layers may be seen. The cells of the epicarp are longitudinally elongated and have completely beaded walls (ends and sides) (Figs. 2.7, 2.40–2.43).

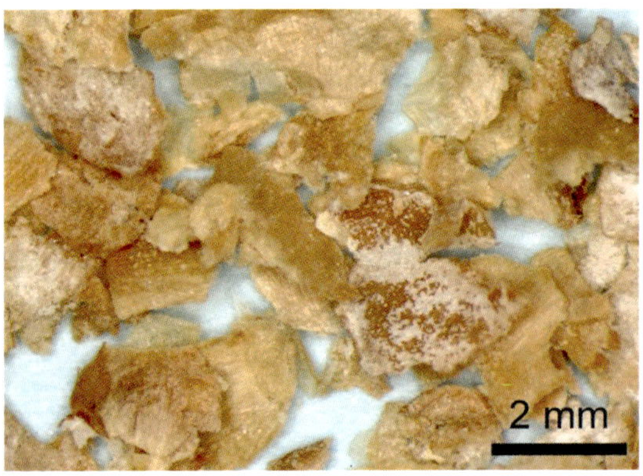

Fig. 2.38. Wheat bran. © Her Majesty the Queen in Right of Canada (2010).

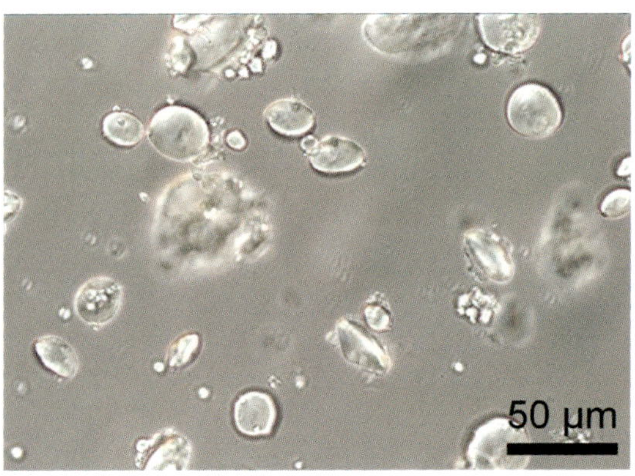

Fig. 2.39. Wheat starch. © Her Majesty the Queen in Right of Canada (2010).

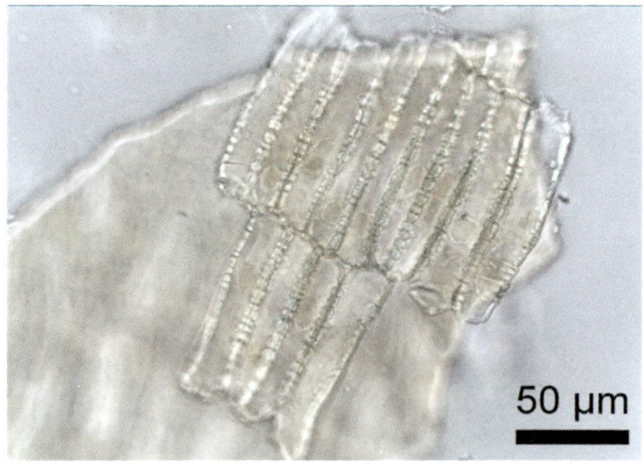

Fig. 2.40. Wheat cross cells. © Her Majesty the Queen in Right of Canada (2010).

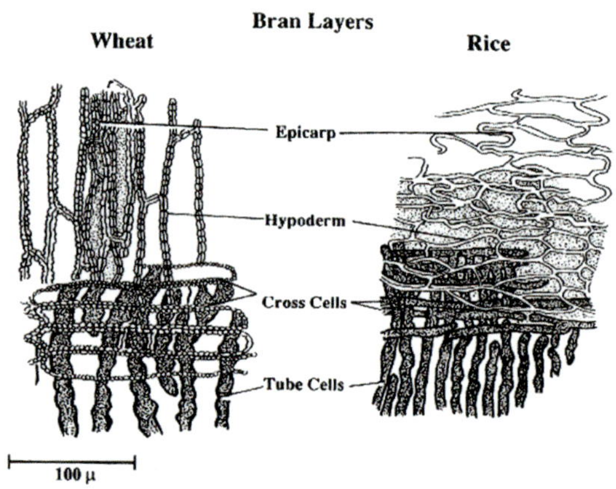

Fig. 2.41. Bran layers.

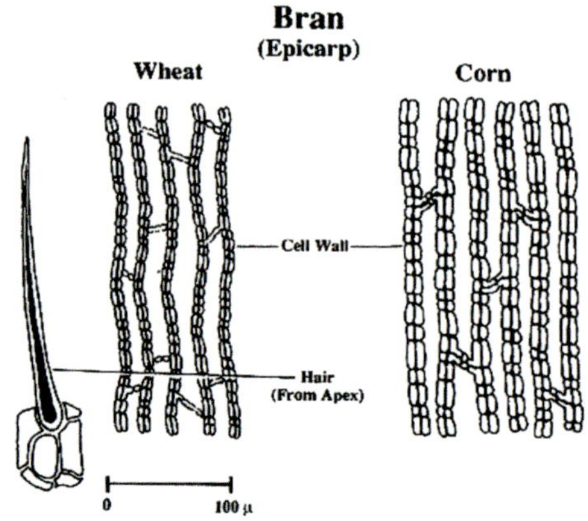

Fig. 2.42. Bran (epicarp).

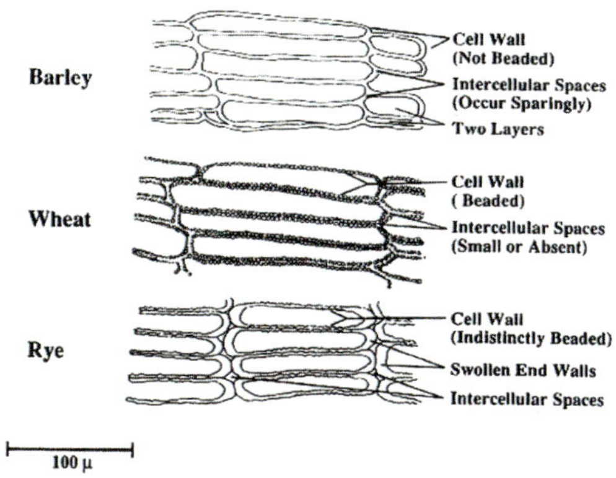

Fig. 2.43. Cross cells.

Oilseeds

Legumes and ***oilseeds*** are, with few exceptions, seeds from dicotyledonous plants. They have two cotyledons, or "seed leaves," and are structurally distinct from cereal grains and grasses that have a single monocotyledon. The term "oilseed" is not a botanical term, but rather one used in industry and commodities. As a result, there can be some overlap between oilseeds and some cereal grains. For example, in some areas of the world, corn, rice, and sorghum may be considered to be oilseeds. Oilseeds include legumes, such as soybeans, as well as some nuts, fruits, and other plants. For the purposes of this manual, oilseeds will refer to seeds with reserve materials primarily in the form of fats or oils.

Canola Meal—*Brassica spp.*

Developed in Canada, canola is a group of selectively bred rapeseed cultivars with low erucic acid and glucosinolate content. Due to the negative connotation of the term "rape" in North America, the plant was renamed canola for ***Can***adian ***o***il, ***l***ow ***a***cid. It is grown both for oil and a high protein meal for feeds.

Definition

Canola meal is the term applied to the selectively bred "double low" varieties of rapeseed containing <5% erucic acid and <20 µmol/g of glucosinolate. *Brassica napus* and *B. campestris* cultivars contributed to the development of canola.

Macroscopic Features

The seeds are generally grayish black to reddish brown. Some new varieties are yellowish. The seeds are round and vary in size from approximately 2 to 3 mm (Fig. 2.44).

The seeds are similar in appearance to wild mustard. The seed can be heated in a few drops of 6 N HCl to help differentiate between canola and wild mustard. The color pigments from the wild mustard will turn the solution red while canola will only slightly discolor the solution a yellow/tan.

Canola meal is light yellow to tan with pieces of hull embedded in the particles (Fig. 2.45).

Fig. 2.44. Whole canola. © Her Majesty the Queen in Right of Canada (2010).

Fig. 2.45. Canola meal. © Her Majesty the Queen in Right of Canada (2010).

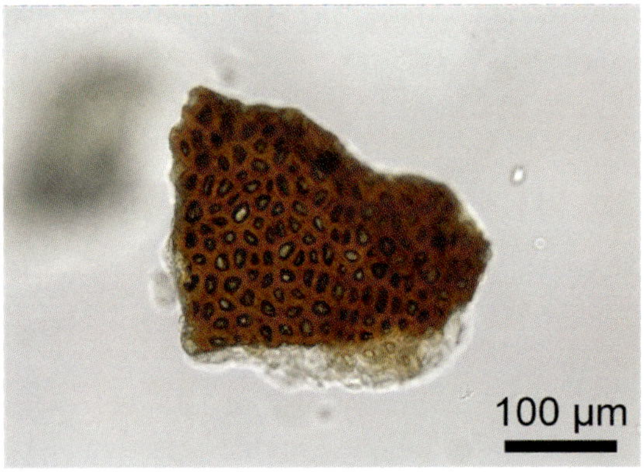

Fig. 2.46. Canola palisade cells. © Her Majesty the Queen in Right of Canada (2010).

Stereoscopic Features

The meal may be identified by the small hull fragments with finely reticulated outer surfaces. The hulls are very obvious and are a distinct diagnostic feature. Many of the hulls are curled. Weed seeds and canola stems are commonly present.

Histological Features

The palisade cells of canola are distinctive. They are four- or five-sided with thick walls and a broad lumen. This lumen, as seen in surface view, is wider than the thick walls. The structure of the thick walls give the cells a honeycomb-like appearance while the color of these cells ranges from yellow to burgundy, with the majority of the cells having an orange-red shade (Fig. 2.46).

Cottonseed Products—*Gossypium hirsutum*

Cottonseed is a major source of natural protein concentrate for cattle feeds; smaller quantities are added to poultry and swine rations. After the cotton fibers are removed from the seed by ginning and delinting, most of the hulls are removed. The kernel is cooked, flaked, and the oil is then extracted by either mechanically or by the use of solvents. After oil removal, the residue is ground into meal and used in feeds and fertilizers.

Definitions

Cottonseed meal, mechanical extraction is the product obtained by grinding the flakes remaining after removal of most of the oil from cottonseed by mechanical extraction.

Cottonseed meal, solvent extraction is the product obtained by grinding the flakes which remain after removal of most of the oil from cottonseed by solvent extraction.

Cottonseed hulls consists primarily of the outer covering of the cottonseed with fibers attached.

Macroscopic Features

Cottonseeds, after removal of the fibers, are pointed, ovoid, dark brow to black, and range in size to 12 mm in length. Seed characteristics and cotton fibers make reliable identifying features (Fig. 2.47).

Cottonseed meal varies from light yellow to golden (if solvent extraction is used) to dark brown (if mechanical extraction is used), depending on the temperature and processing time. Solvent extracted meal contains less oil than mechanically extracted meal (Fig. 2.48).

Stereoscopic Features

Hull fragments are brown to dark brown, relatively thick (up to 400 μm), tough, and leathery. Variegated colored layers are visible on the hull edges. Large particles of hull show the rounded seed contour. Twisted cotton

Fig. 2.47. Whole cottonseed. © Her Majesty the Queen in Right of Canada (2010).

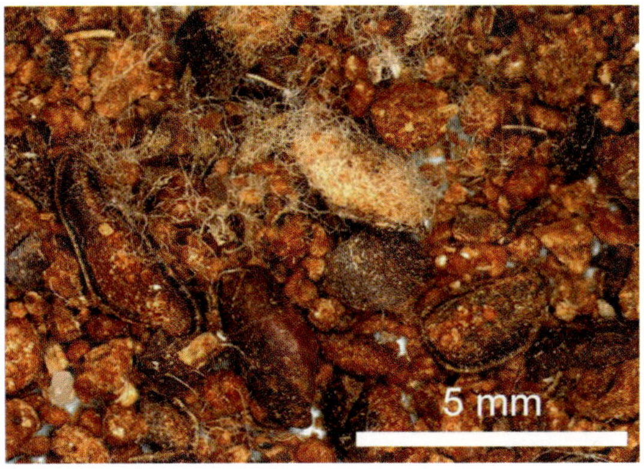

Fig. 2.48. Cottonseed meal. © Her Majesty the Queen in Right of Canada (2010).

fibers are usually found adhering to the hulls and embedded in chunks of the meal. Characteristic reddish brown pigment glands are visible in raw ground cottonseeds and many ground meals. (Figs. 2.49 and 2.50).

Histological Features

The hull fragments are deep red-orange color with ovoid sinuses that appear as darker areas in a rosette pattern (Fig. 2.51). The cotton fibers appear as flattened and twisted bands that may bend at severe angles onto themselves; the walls are much thinner than the lumen. These features distinguish cotton fibers from hair, wool, and synthetic fibers (Fig. 2.52). Cottonseed meal particles appear as greenish yellow masses of distorted cells, some with oil glands.

Fig. 2.49. Cottonseed hulls. © Her Majesty the Queen in Right of Canada (2010).

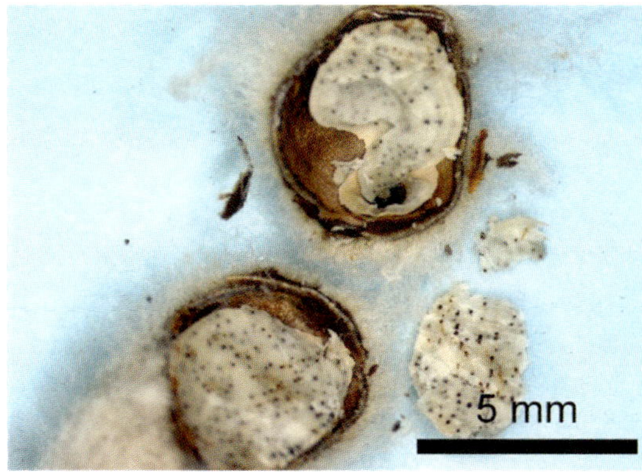

Fig. 2.50. Cottonseed meal gossypol. © Her Majesty the Queen in Right of Canada (2010).

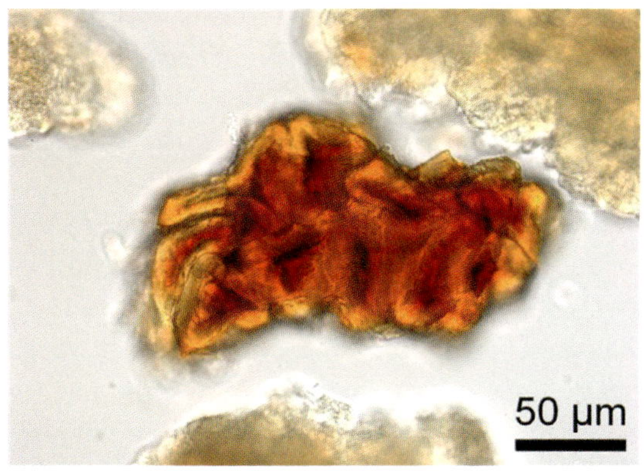

Fig. 2.51. Cottonseed meal—outer epiderm. © Her Majesty the Queen in Right of Canada (2010).

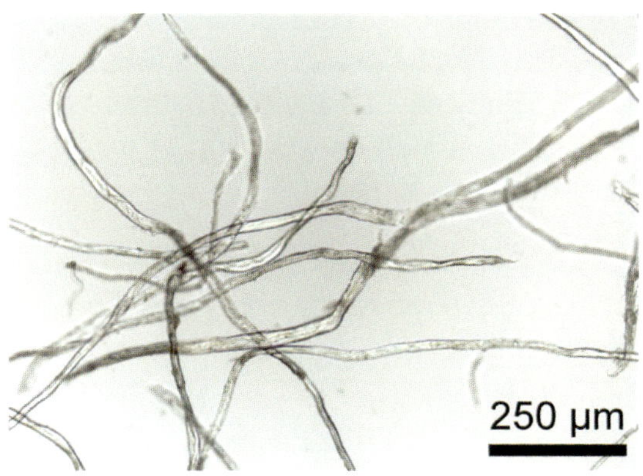

Fig. 2.52. Cotton fibers. © Her Majesty the Queen in Right of Canada (2010).

Flaxseed (Linseed) Products—*Linum usitatissimum* L.

The cultivation of flax for fiber dates back to early civilization. Linseed meal or cake has long been used in animal feeds. This meal is manufactured by either pressing the crushed and heated flaxseed or by using solvents to remove the oil.

Fig. 2.53. Whole linseed. © Her Majesty the Queen in Right of Canada (2010).

Definitions

Linseed meal, mechanical extraction is the product obtained by grinding the cake or chips after removal of most of the oil from flaxseed by mechanical extraction.

Linseed meal, solvent extraction is the product obtained by grinding the cake or chips after removal of most of the oil from flaxseed by solvent extraction.

Flaxseed screenings meal is the ground product obtained after solvent extraction of part of the oil from the smaller, imperfect flaxseed. This meal typically contains weed seeds and other foreign materials having some feeding value that remain after cleaning flaxseed.

Flax plant product is that portion of the flax plant which remains after harvesting the seed and separating the bast fibers and flax shives. Flax plant product consists of the leaves, cortical tissues, flaxseed bolls, and broken and immature flaxseed.

Flax straw by-product is the ground product remaining after the removal of the longer fiber material from the flax plant by mechanical processing.

Macroscopic Features

Flaxseed is a flattened, ellipsoidal, lustrous, smooth, brown or yellow seed, ranging in size to 6 mm (Fig. 2.53). Its high oil content can be demonstrated by applying pressure to the seeds. Linseed products are brown to reddish brown. The presence of weed seeds and other contaminants alters the color of flaxseed screenings. Flax plant product is light brown and fibrous with gray-white stem particles. If oil is present, flax products have a characteristic odor.

Stereoscopic Features

Processed flaxseed products are recognized by the small particles of thin, light to dark brown papery hull or outer covering. Red pigment particles embedded in the white meal give it a speckled appearance. This meal usually remains attached to the hull. Flax plant product may be identified by the presence of unextracted flaxseed with its high oil content and plant parts. Weed seeds and sand are often found in flaxseed screenings meal and flax plant product.

Histological Features

The chief characteristic of linseed meal is the pigmented testa cells. These cells are rectangular, often square, with finely beaded walls and dark red contents. Intact testa sections may be found or the pigment cells may rupture, releasing the pigmented contents to float free in the mounting medium (Fig. 2.54). The fiber cells of the hull are also characteristic, being arranged in single or double layers and having thick walls.

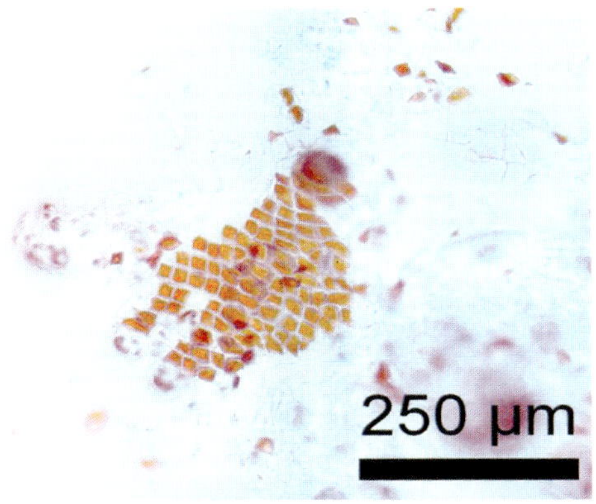

Fig. 2.54. Linseed testa cells. © Her Majesty the Queen in Right of Canada (2010).

Peanut Products—*Arachis hypogaea L.*

Peanut products are primarily used as food, but the concentrate, peanut meal, and the "roughage" peanut hulls, are used in animal feeds. Other products such as peanut skins, vine meal, and stem meal, are also used in commercial feeds.

Definitions

Peanut meal and hulls is a product comprised of shelled peanuts, primarily the hulls and kernels, from which most of the oil has been removed by mechanical or solvent extraction.

Peanut meal is a ground product comprised primarily of ground peanut kernels, containing hulls, fiber and/or oil as typically remains after normal manufacturing processes.

Peanut skins is the outer covering of the peanut kernel exclusive of hulls as obtained in ordinary commercial processing. This product may contain broken peanut kernels.

Peanut hulls consist of the outer hull of the peanut shell.

Ground peanut hay is composed of ground peanut leaves and stems from which the peanuts have been removed.

Macroscopic Features

Peanuts are pods commonly containing two kernels, less often one or three, and rarely four or five. The pod shells, or hulls, are brittle and vary from a light buff color to yellow or orange. The outer hull surface has longitudinal and lateral ridges interwoven to form a network. The inner hull has a soft, white, silky surface. The seeds are crowded during development, so their ends are flattened, and each seed is covered with a thin, copper to red skin (testa). The cotyledons are semi-circular, yellow or buff, oily, and have a longitudinal groove through the middle of their inner surface (Fig. 2.55).

Stereoscopic Features

The hulls and red skins are the most characteristic features of the peanut, and their fragments are usually present in peanut products. (Figs. 2.56 and 2.57). The ridges and reticulations of the hull are valuable for identification. The dense fiber strands on the hull surface and the silky inner hull are often broken from the hull during grinding but remain identifiable. Even fairly small hull fragments are thick, allowing peanut hulls to be distinguished from wheat middlings. The skin is reddish brown, copper or purple on the outside, and deep yellow to white on the inside. The ground seed are generally buff, but may be brown to creamy white after roasting.

Fig. 2.55. Whole peanuts.

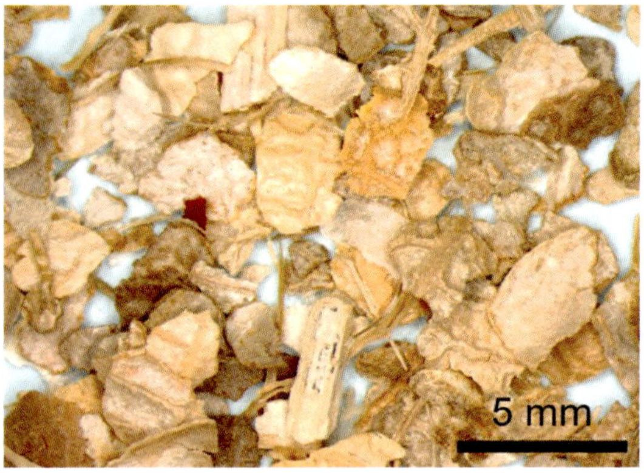

Fig. 2.56. Peanut hulls and skins. © Her Majesty the Queen in Right of Canada (2010).

Fig. 2.57. Peanut skins. © Her Majesty the Queen in Right of Canada (2010).

Some pink discoloration may be seen in meal fragments. The lack of distinctive features and processing variation can make peanut meal difficult to identify.

Stem meal, vine meal and peanut hay are much coarser than other peanut products. The woody stem fragments are conspicuous; they have dark, longitudinal, resinous deposits on the inside portions. Leaves and leaf fragments are often found, particularly in vine meal. The leaves are distinctly veined, and the margins are covered with long, slender, twisted hairs or fibers. These hairs can be difficult to find; they are often broken or rubbed off during grinding, cleaning, or handling.

Sand particles are a common contaminant.

Histological Features

The shell (achene) has an outer layer of thin-walled cells; many have outgrowths similar to root hairs. The hypodermal region consists of several layers of radially arranged cells—rectangular, thin-walled outer ones, and thick-walled, pitted inner ones. The mesocarp is characteristically fibrous, with projections into the achene. Crisscrossing of fibers at all angles is valuable for identification of shell fragments.

The outer epidermis of the skin (testa) lacks palisade cells. Testa cells are four- and five-sided with thickened and pitted outer walls that resemble a double row of pennants back-to-back. At 200×, skin fragments appear splotched with red or yellow color.

The cotyledon is composed of parenchymal cells with pitted, thick walls. Cotyledon cells contain small starch granules in addition to oil and aleurone grains. These starch granules are small, biconcave discs, resembling red blood cells.

Peanut leaf trichomes are long and pointed, approximately 15 μm in diameter, and the lumen is thicker than the wall.

Peanut starch is rounded and has a central depression, making it appear similar to that of a red blood cell in that it resembles a biconcave disc (Fig. 2.58).

Safflower Products—*Carthamus tinctorius*

Safflower meal is one of the world's oldest crops, grown in India as a source of oil for more than 150 years. Safflower is used both as a food and industrial quality oil source. The residual meal is used for livestock and poultry.

Definitions

Safflower meal, mechanical extracted is the ground residue obtained after extracting the oil from whole safflower seed by mechanical means.

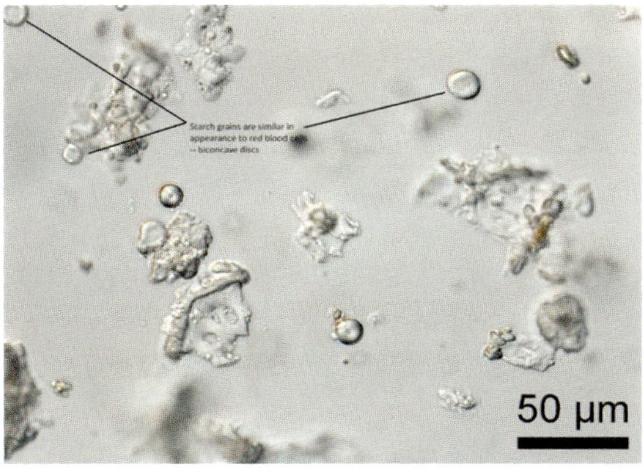

Fig. 2.58. Peanut starch. © Her Majesty the Queen in Right of Canada (2010).

Fig. 2.59. Whole safflower. © Her Majesty the Queen in Right of Canada (2010).

Safflower meal, solvent extracted is the ground residue obtained after extracting the oil from whole safflower seed by solvent extraction.

Macroscopic Features

Safflower seeds are irregularly pear-shaped, approximately 1 cm long, with a smooth, shiny surface. It is pale gray to white in color. The meal is light colored to tan, and always contains hull particles (Fig. 2.59).

Fig. 2.60. Safflower seed meal. © Her Majesty the Queen in Right of Canada (2010).

Stereoscopic Features

The hull is characterized by dark reddish to black phytomelanin pigments sandwiched between white or light-colored layers. This pigmented layer is very fine and observed only along the edge of the hull particles (Fig. 2.60).

Histological Features

All hull layers (pericarp, inner pericarp, and endocarp) contain stone cells. The meal may be recognized by the numerous stone cells, the dark reddish to black phytomelanin layer, and the pitted cells of the hull (Fig. 2.61).

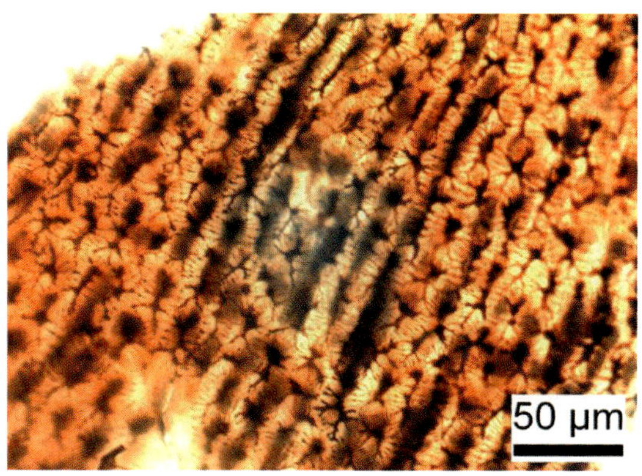

Fig. 2.61. Phytomelanin cells. © Her Majesty the Queen in Right of Canada (2010).

Soybean Products—*Glycine max*

Soybean products are used in virtually all types of feeds. Soybean meal (extracted or full-fat) and protein concentrates or isolates are extensively utilized protein sources. Although unprocessed soybeans are used for feed, most soybeans are processed to minimize anti-nutritional factors. All soybean feed products must be labeled according to the method of manufacture.

Definitions

Ground soybeans is obtained by grinding whole soybeans without cooking or removing any of the oil.

Ground soybean hay is the ground soybean plant including the leaves and beans. It must be reasonably free of other crop plants and weeds.

Soybean hulls consist primarily of the outer covering of the soybean.

Soybean meal, mechanical extracted is the product obtained by grinding the cake or chips remaining after removal of most of the oil from soybeans by mechanical extraction. It may contain an inert, non-toxic conditioning agent to reduce caking and improve its flow in an amount not to exceed that necessary to accomplish its intended effect and in no case to exceed 0.5%.

Soybean meal, solvent extracted is the product obtained by grinding the flakes remaining after removal of most of the oil from soybeans by solvent extraction. It may contain an inert, non-toxic conditioning agent as described for mechanical extracted soybean meal.

Soybean meal, dehulled, solvent extracted is obtained by grinding the flakes remaining after removal of most of the oil from dehulled soybeans by solvent extraction. Heat must be applied during the process. It may contain a conditioning agent as described for mechanical extracted soybean meal.

Soybean mill feed is composed of soybean hulls and the offal from the tail of the mill which results from the manufacture of soy grits or flour.

Soybean mill run is composed of soybean hulls and such bean meats that adhere to the hulls which results from normal milling operations during the production of dehulled soybean meal.

Heat processed soybeans is the product resulting from heating whole soybeans without removing any of the component parts. It may be ground, pelleted, flaked, or powdered.

Ground extruded whole soybeans is the meal product resulting from the extrusion by friction, heat, and/or steam of whole soybeans without removing any of the component parts.

Kibbled soybean meal is the product obtained by cooking ground solvent extracted soybean meal under pressure and extruding from an expeller or other mechanical pressure device.

Soy protein concentrate is prepared from high quality, sound, clean, dehulled soybean seeds by removing most of the oil and water soluble nonprotein constituents.

Soy grits is the granular material resulting from the screened and graded product after removal of most of the oil from selected sound, clean, and dehulled soybeans by a mechanical or solvent extraction process. Both the finely powdered material resulting from this process and whole, ground, raw soybeans may be called ***soy flour.***

Macroscopic Features

The soybean is a typical legume. The hairy pods range from 4 to 7 cm long and usually contain two to four seeds. The seeds are generally elliptical, very slightly flattened, and approximately 6 to 10 mm in diameter. The hull or seed coat is smooth, shiny, and firm in texture. Most American commercial cultivars are yellow, although they may range from green through buff to black. The hilum, or seed scar, is ovoid in shape, approximately 3 mm × 1 mm (Fig. 2.62).

Stereoscopic Features

Most soybean products can be identified at low magnification by fragments of hull. Even dehulled soybean products will have a small percentage of hull fragments present.

Fig. 2.62. Soybean in pod. © Her Majesty the Queen in Right of Canada (2010).

The oval shaped hilum is commonly black or dark brown but may range from buff to almost colorless. The slit and the edges of the hilum are distinct, raised, and lighter in color than the rest of the hilum (Fig. 2.63).

The outer surface of the hull is uniquely pockmarked giving the hull an appearance akin to that of a citrus fruit skin. The inner surface if off white and has a slightly filamentous, porous, spongy appearance. The entire hull is approximately 0.1 mm thick (Fig. 2.64).

Soybean meal consists of particles of the embryo from which most of the oil has been removed. It may have hulls added back to adjust the protein content. Added hulls tend to be flat while soybean meal hulls tend to be curled.

Particles of solvent extracted meal are irregular and flat with rounded edges. The particles are translucent with a glazed surface and vary from cream to pale reddish brown. Expeller meal is usually darker and forms rough, granular chunks rather than rounded, flattened particles.

Fig. 2.63. Soybean meal. © Her Majesty the Queen in Right of Canada (2010).

Fig. 2.64. Soybean hulls. © Her Majesty the Queen in Right of Canada (2010).

Histological Features

The palisade cells and sub-epidermal cells (also referred to as "hourglass" cells due to their shape) of the hull are the most useful structures for histological identification (Figs. 2.65 and 2.66).

In finely ground samples, the palisade cells are often floating free in the medium or associated in groups with or without attached "hourglass" cells. The palisade cells range from 40 to 60 μm long and are much thinner in diameter. The sub-epidermal "hourglass" cells are hexagonal from an end view and vary in

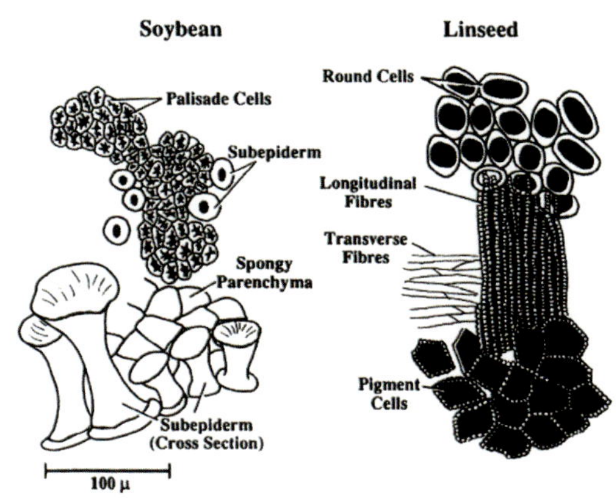

Fig. 2.65. Soybean linseed.

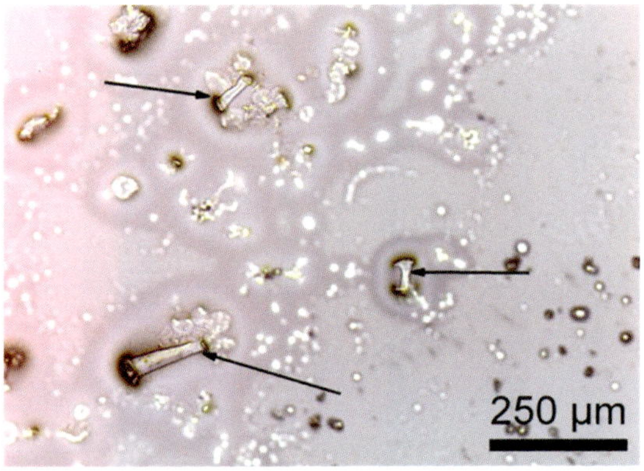

Fig. 2.66. Soybean hour glass cells. © Her Majesty the Queen in Right of Canada (2010).

length from 30 to 70 µm. They may be found singly or associated with palisade cells.

Intact hull pieces, when viewed from the outer surface, will appear as black, rough-edged, cracked, and irregular polka dots on a colorless to tan translucent background. Hull pieces are relatively dense and may be found in the sediment fraction if a sample is floated using chloroform. These pieces may be misidentified as bone fragments when viewed from the surface. Observing the cells using polarized light helps distinguish between bone fragments and palisade cells as the palisade cells are much more brightly birefringent than bone fragments, as are the "hourglass" cells (Fig. 2.67).

Sunflower Products—*Helianthus annus L.*

Definitions

Sunflower meal, mechanical extracted is obtained by grinding the residue remaining after extraction of the oil from the whole sunflower seed by mechanical extraction.

Sunflower meal, solvent extraction is obtained by grinding the residue remaining after extraction of the oil from the whole sunflower seed by solvent extraction.

Sunflower hulls consists of the outer covering of sunflower seed.

Sunflower meal, dehulled, mechanical extracted is obtained by grinding the residue remaining after the mechanical extraction process and from which most of the hulls are removed.

Sunflower meal, dehulled, solvent extracted is obtained by grinding the residue remaining after the solvent extraction process and from which most of the hulls have been removed.

Macroscopic Features

The seeds are obovate, up to 18 mm long, flattened, and diamond-shaped in cross section. They are black or black and white striped (Fig. 2.68).

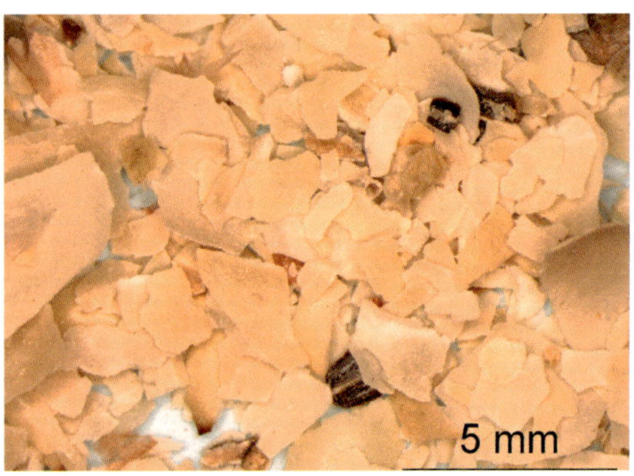

Fig. 2.67. Soybean hulls. © Her Majesty the Queen in Right of Canada (2010).

Fig. 2.68. Sunflower meal. © Her Majesty the Queen in Right of Canada (2010).

Stereoscopic Features

The meal is light grey to dark greenish grey. Hull fragments are always present. These fragments are black or black and white striped. The meal granules appear speckled due to imbedded hull fragments (Fig. 2.69).

Hulls are leathery, with a paper-like cream lining. They are the key to identification of sunflower products.

Histological Features

The outer epidermal layer contains highly characteristic transversely elongated cells with zigzag walls. A thickening is seen at each angle of the zigzag, sometimes forming a loop.

Sunflower meal fines can be identified by twin hairs and the black phytomelanin pigment layer. (Figs. 2.70 and 2.71).

Fig. 2.69. Whole sunflower oil seeds. © Her Majesty the Queen in Right of Canada (2010).

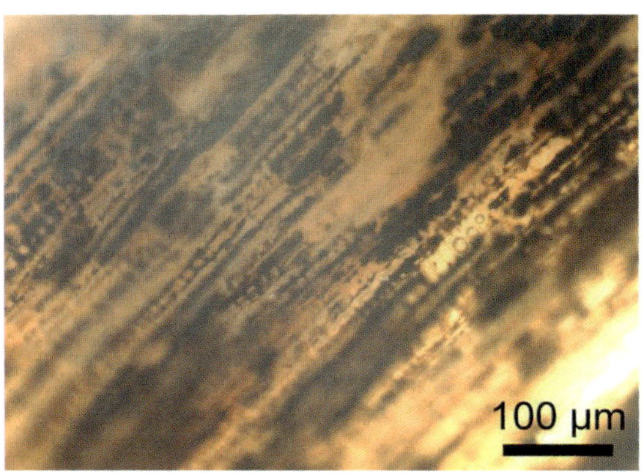

Fig. 2.70. Phytomelanin. © Her Majesty the Queen in Right of Canada (2010).

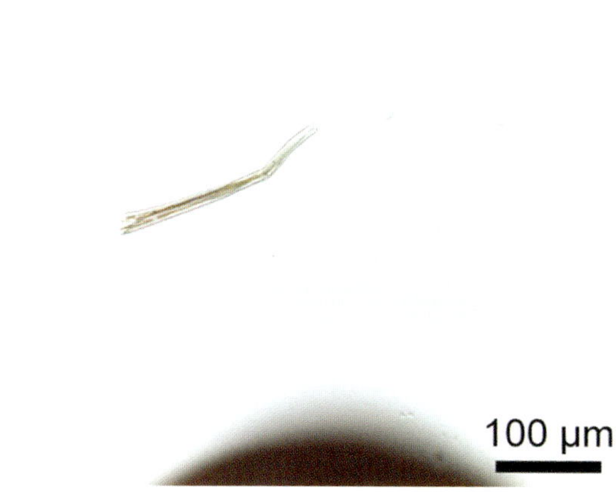

Fig. 2.71. Twin hairs. © Her Majesty the Queen in Right of Canada (2010).

Miscellaneous Plant Materials

The remaining agricultural products, though important in agricultural and agrifood products, do not fit neatly into a particular category, such as cereal grains or oilseeds.

Alfalfa Products—*Medicago sativa*

Alfalfa is a leguminous forage and fodder plant. Alfalfa is extensively grown as a green forage, hay, and as a component of manufactured feed products. Gerbil and rabbit rations may contain large quantities of alfalfa products.

Alfalfa (also called Lucerne) can be infested with toxic blister beetles (three species: black, gray, and/or yellow with brown stripes).

Definitions

Sun cured alfalfa meal, pellets, or ground alfalfa hay is the aerial portion of the alfalfa plant which has been sun cured and chopped or finely ground. It must be reasonably free of other crop plants, weeds, and mold.

Dehydrated alfalfa meal or pellets is the aerial portion of the alfalfa plant which has been thermally dried and finely ground or pelleted. It must be reasonably free of other crop plants, weeds, and mold.

Macroscopic Features

Alfalfa products vary from bright green to brown or grey, depending on the degree of maturity during cutting, the drying method used, and the storage conditions. Freshly prepared dehydrated products are a bright green; they fade to a grey-brown on exposure to light and finally turn grey or grey-brown. Brown-colored particles in alfalfa meal are usually mature stems but can also be browned leaf tissue. Grass hay or meal becomes brown when faded and may be included in poor quality alfalfa products. Humus, which can be present in low-grade products, also causes a brown color. Charred alfalfa particles may be dark brown to black. (Figs. 2.72 and 2.73).

A green color remaining on the filter paper after exposure to solvents during flotation (chlorophyll), suggests the presence of alfalfa.

Stereoscopic Features

Coarsely ground alfalfa products are identified by the stem, leaves, flowers, and developing seeds. The stem is smooth and semi-woody. The leaves are oblong and sharply serrated or toothed on the upper one-third of the margin. The veins occur semi-longitudinally along the central rib. The tip of the leaf is terminated by the projected midrib.

The flowers of alfalfa are purple, white, yellow, blue, green, or variegated. The fruit is a spirally twisted pod containing from one to eight small, mitten-shaped seeds. The seeds (up to 3 mm in diameter) are olive-green to greenish brown.

Fig. 2.72. Alfalfa meal. © Her Majesty the Queen in Right of Canada (2010).

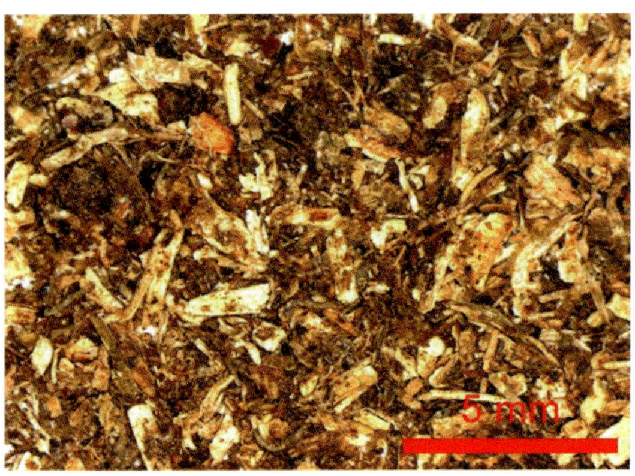

Fig. 2.73. Alfalfa meal (brown, faded). © Her Majesty the Queen in Right of Canada (2010).

Histological Features

In the leaf, cell walls of the lower epiderm are wavy in outline and contain trichomes (hairs). These trichomes are approximately 15 µm in diameter with distinct warts along their entire length. Long trichomes with warts distinguish alfalfa products from red and alsike clovers (Fig. 2.74).

The parallel veining, size, shape, and linear arrangement of stomata distinguish the leaf tissues of monocots (grasses) from dicots.

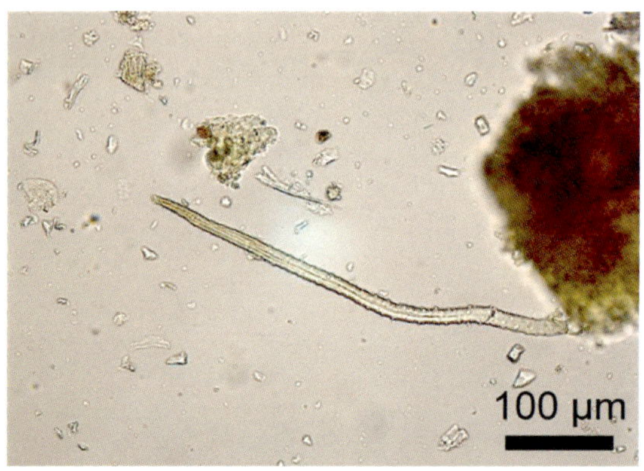

Fig. 2.74. Alfalfa trichome. © Her Majesty the Queen in Right of Canada (2010).

Brewers Products

Brewers products are cereal grain residues from beer and malt liquor production. They are used primarily in ruminant feeds.

Definitions

Brewers dried grains is the dried, extracted residue of barley malt, alone or in mixture with other cereal grains or grain products resulting from the manufacture of wort or beer. It may contain pulverized, dried, spent hops in amounts not exceeding AAFCO definition.

Malt sprouts is the product obtained from malted barley by removal of the rootlets and sprouts and may contain malt (barley) hulls, other parts of malt, and foreign material unavoidably present.

Malt cleanings is obtained from cleaning malted (germinated or enzyme-activated) barley or from re-cleaning malt which has been deemed unacceptable.

Macroscopic Features

Brewers dried grains are characterized by the crushed, twisted, and broken hulls of barley with the steel gray colored endosperm residues. Structural characteristics of barley remain for identification; some of the broken barley hulls will retain the triangular shape of ground barley. The malting process leaves the barley hull lusterless and grayish tan. Other starchy adjuncts found in brewers products, such as rice, corn, and grain sorghum, will resemble the products of wet milling (Fig. 2.75).

Malt sprouts are the broken radicles (rootlets) of sprouted barley. They are light yellow to dark tan, wrinkled, rod-shaped, and vary from 0.25 to 0.30 mm in diameter. They will appear rounded on one end and blunt on the other (the rounded end is the growth end while the blunt end is the end having been cut). Their length may range up to an inch or more (Fig. 2.76).

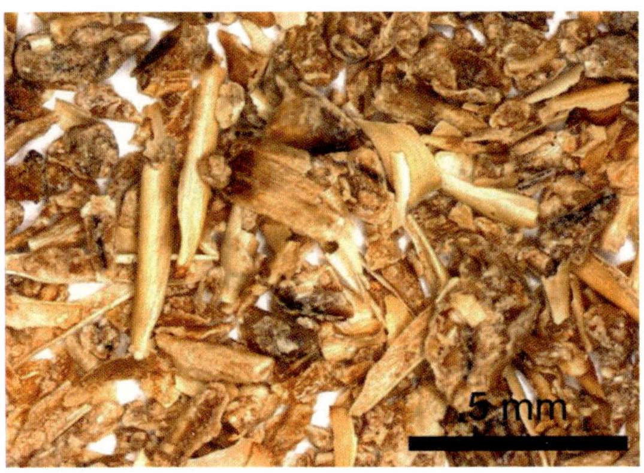

Fig. 2.75. Brewers dried grains. © Her Majesty the Queen in Right of Canada (2010).

Fig. 2.76. Malt sprouts. © Her Majesty the Queen in Right of Canada (2010).

Stereoscopic Features

The structural characteristics of barley hull and other grains are not destroyed during brewing. The barley hulls will typically appear to be longitudinally curved, filled with smaller particles of fine material. Malt sprouts are characterized by a rod-shaped structure, a yellow to dark tan color, and longitudinal wrinkles. The sprouts may be straight to curved small rods, blunt on one end and rounded on the other.

Histological Features

Histological features are not required to identify brewers by-products; most of them are detectable under low magnification. The wavy-walled cells of barley hulls and the grayish masses of fine material inside the longitudinally curved barley hulls are diagnostic features. Occasionally, starch granules are present.

Buckwheat Feed

Buckwheat feed is a mixture of buckwheat middlings and hulls. This product can be identified by the very prominent black pieces of pointed triangular hull with thick-walled elongated cells and the spongy parenchyma of the bran layer.

Citrus Products

Citrus products are the by-products of the orange and grapefruit industry and consist of the dried residues following juicing and pulping.

Definitions

Dried citrus pulp is the ground peel, residue of the inside portions, and occasional cull fruits of the citrus family which have been dried, producing a coarse, thickly flaked product. It may contain dried citrus mean and citrus seeds.

Dried citrus meal consists of the finer particles derived from screening dried citrus pulp.

Citrus seed meal is a product obtained from the seed or seed meats of oranges and grapefruit from which most of the oil has been removed. It is composed mostly of the kernel with portions of the hull and pulp that were unable to be removed during the manufacture of citrus seed oil.

Macroscopic Features

The peel, seed, and odor are perhaps the most useful features in citrus pulp identification. Pieces of peel become twisted and darkened during processing/drying. Peels may be yellow, green, orange, or brown, depending on the original color and processing. The outer surface of the peel is pebbled with oil glands. The inner surface of the peel is continuous with the outer surface and is generally cream to tan colored.

In citrus pulp, the seeds are usually whole. Seeds range in size from 10 to 15 mm and are oblong, with the seed coat extending beyond the endosperm on both ends to either a point or wedge. The seed coat is buff colored and often wrinkled or ridged.

Pieces of semitransparent buff to gray membrane from the walls of the fruit sections, sting-like vascular bundles which form a net over the fruit and attachment buttons, may also be present.

Citrus meal consists of small, delicate particles of membrane and vascular bundles. Small amounts of peel and seed fragments may be present. Since these particles are present after screening, they will be small in size. Some of the particles may be charred because they become airborne during drying and quickly overheat. (Figs. 2.77 and 2.78).

Stereoscopic Features

Citrus meal is recognized by the presence of twisted particles of the peel and membrane. Citrus seed meal is identified by the particles of the seed coat. The inner seed coat is a brown semitransparent membrane intensely colored at the free end of the seed. Inside the seed coat is the oily, yellow-to-green endosperm.

Fig. 2.77. Citrus meal. © Her Majesty the Queen in Right of Canada (2010).

Fig. 2.78. Coarse citrus meal. © Her Majesty the Queen in Right of Canada (2010).

Histological Features

Histological features are not generally used for the identification of citrus products.

Coconut Meal—*Cocos nucifera*

Coconut, or copra meal, is the ground residue obtained after extraction of part of the oil from dried coconut meat.

Definitions

Coconut meal, mechanical extraction is the ground residue which remains after removal of most of the oil from dried coconut meat by mechanical extraction.

Coconut meal, solvent extraction is the ground residue which remains after removal of most of the oil from dried coconut meat by solvent extraction.

Macroscopic Features

Coconut meal should be whitish to medium brown. During oil extraction, excessive temperature will result in a darker meal of reduced quality with decreased digestibility. Black, irregularly shaped particles from the shell can be identified. Coconut meal has a distinct, pleasant odor which indicates its presence in a mixed feed.

Stereoscopic Features

Coconut meal is characterized by irregularly shaped, flaky fragments, varying in color from tan to pale pink or white. They are soft and spongy and may express oil under pressure. The meal also contains black, hard particles of shell, and may contain twisted, woody fibers ranging in color from yellowish brown to brown (Fig. 2.79).

Histological Features

Coconut meal endosperm has large, colorless, straight, thin-walled cells with clear oil vesicles. The ground shell contains stone cells. These cells are yellowish brown and elongated with a dark center.

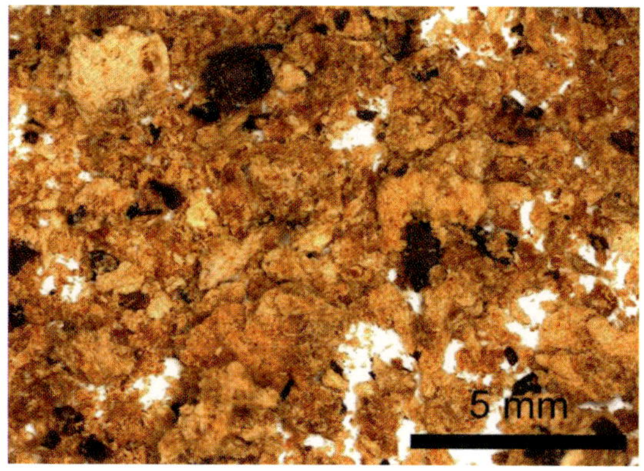

Fig. 2.79. Coconut meal. © Her Majesty the Queen in Right of Canada (2010).

Distillers Products

Distillers products are obtained from the manufacture of distilled liquors and alcohol and may be composed of one or a mixture of grains. They are also obtained from molasses and potatoes. Corn is the most commonly used grain.

Definitions (Using Corn as an Example)

Corn distillers dried grains is the product obtained in the manufacture of distilled liquors and alcohol from corn, or from a grain mixture in which corn predominates, by drying that portion of the whole stillage retained by screens.

Corn distillers dried solubles is the product obtained in the manufacture of distilled liquors and alcohol from corn, or from a grain mixture in which corn predominates, by condensing and drying the stillage fraction which passes through the screens. If at least three-fourths of the solids from the whole stillage are dried with the solubles, the resulting product is called ***corn sistillers dried grains with solubles.***

Other distillers dried grains and solubles are labeled to identify the grain source which predominates (i.e., ***barley dried grains***).

Molasses distillers dried solubles is the product obtained by drying the residue from the yeast fermentation of molasses after removal of the alcohol by distillation.

Potato distillers dried residue is the dried product obtained after the manufacture of alcohol and distilled liquors from potatoes or from a mixture in which potatoes predominate.

Macroscopic Features

Distillers dried grains are somewhat similar in appearance to gluten feeds but are usually darker in color and contain little or none of the starchy portions of the grain. Distillers solubles are red to reddish-brown in color and have a flaky consistency, a sharp odor, and a sour taste. The odor itself may be proof of the presence of this product in a mixed feed. Molasses distillers dried solubles are black, shiny, granulated, or flaky meal, with numerous chunks or aggregates due to moisture absorption.

Stereoscopic Features

Distillers grains are characterized by the bran and hulls of the grains used in the distilling process. Corn bran particles separate readily from the remaining endosperm cells. These particles may be flat, slightly curled, or bent, rather than rolled as in wheat, barley and grain sorghum Distillers grains. The linear bran markings are more prominent than on unprocessed corn bran. Small amounts of the ground cleanings may be seen.

Distillers solubles occur as irregularly shaped reddish-brown flakes and are sticky when pressed. If the solubles are drum dried, one surface is relatively smooth, the other very irregular or rough. Upon close examination, small holes may be found. With direct light, small crystalline surfaces may be seen.

Molasses distillers dried solubles are purplish black aggregates that look shiny, as if wet. They have rough surfaces and are easily crushed into powder with forceps. In water they disintegrate slowly, with color diffusing out rather rapidly.

Histological Features

The structures of the bran, trichomes, and occasional starch granules of the processed grains may be used as diagnostic features. Bran is most abundant and is usually conclusive in identifying distillers grains. Masses of the endosperm may be found and appear as darkened globules with lighter colored margins.

Distillers solubles are amorphous, reddish brown, opaque flakes. From the surface, oblong openings of variable size can be seen. A flake may bear dark and light areas due to varying thicknesses. An occasional small piece of bran may also be observed. (Figs. 2.80, 2.81, 2.82, and 2.83).

Yeast cells remaining from the fermentation process may be evident.

Fig. 2.80. Corn Distillers with soluble. © Her Majesty the Queen in Right of Canada (2010).

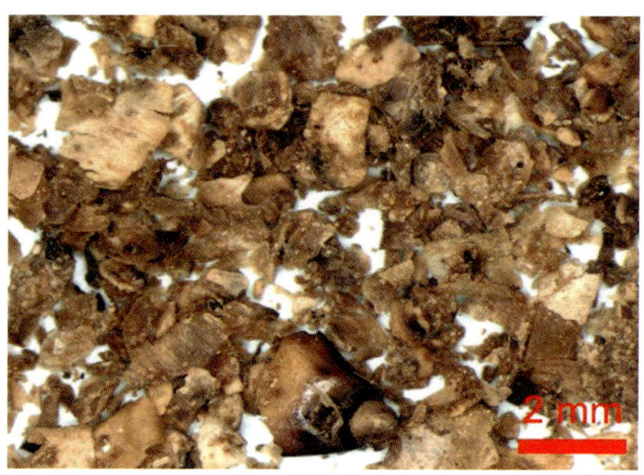

Fig. 2.81. Sorghum (milo) Distillers dried grains. © Her Majesty the Queen in Right of Canada (2010).

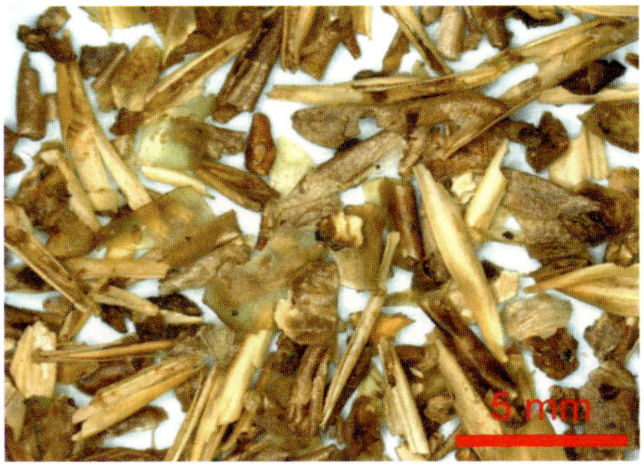

Fig. 2.82. Rye Distillers dried grains. © Her Majesty the Queen in Right of Canada (2010).

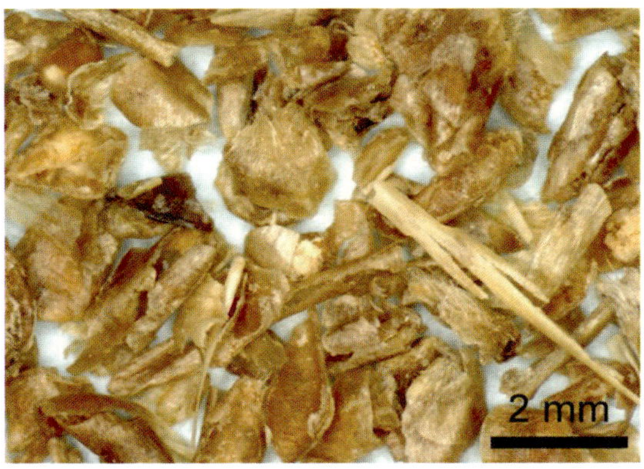

Fig. 2.83. Wheat Distillers dried grains. © Her Majesty the Queen in Right of Canada (2010).

Dried Bakery Product

Dried bakery product is a mixture of bread, cookies, cake, crackers, flours, and dough that have been mechanically separated from non-edible material, artificially dried, and ground. Dried bakery product is often found in hog feeds as a means to enhance palatability.

Some useful keys to identifying this variable by-product are: bright, artificially colored particles, exotic seeds used in bakery products (sesame, caraway, poppy, etc.), and fully gelatinized starches which have some chip fragments (tortilla, potato, etc.) or identifiable crumb structure. The presence of sucrose or other reducing sugars at higher than expected levels is another indication of bakery product possibly being present.

Dried Beet Pulp

Dried beet pulp is the residue from sugar beet processing for sugar. Sugar beets are generally, shredded, not ground, before extraction and then dried for the feed trade. The tough, gray, stringy, slightly curled shreds are very characteristic and distinctive. Beet pulp absorbs water readily and will become soft and swell to several times its original size (Fig. 2.84).

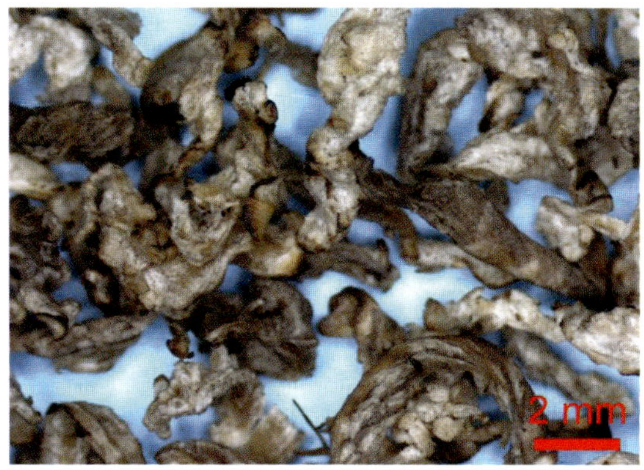

Fig. 2.84. Beet pulp. © Her Majesty the Queen in Right of Canada (2010).

Kelp Meal

Kelp meal is a ground seaweed product used mainly as a source of iodine. Macroscopically, it looks like a flaky, heavy soil. Stereoscopically, the flat, thin particles are generally velvety brown; but color can vary from buff to dark green (Fig. 2.85).

Fig. 2.85. Kelp meal. © Her Majesty the Queen in Right of Canada (2010).

Molasses Products

About 90% of molasses that is fed to livestock is used in the manufacture of formula feeds or commercial mixed feeds. Molasses is produced from sugar cane, sugar beets, corn, and citrus juice and is available in liquid and dehydrated forms. Molasses is used in feeds to increase palatability, to reduce dust, to serve as an energy source, and in some instances, to act as a carrier for certain minerals or other feed ingredients.

Definitions

Beet molasses is a by-product of the manufacture of sucrose from sugar beets.

Cane molasses or blackstrap is a by-product of the manufacture of sucrose from sugar cane.

Citrus molasses is the partially dehydrated juices obtained from the manufacture of dried citrus pulp.

Starch molasses is a by-product of the manufacture of dextrose by hydrolysis of corn or grain sorghum starches.

Hemicellulose extract or wood molasses is a by-product of the manufacture of pressed wood. It contains pentose and hexose sugars.

Macroscopic Features

Most formula feeds containing molasses have the product added in its natural state; a thick, heavy, light yellow to black syrup with a distinct aroma. Citrus molasses has a citrus odor, an exceedingly bitter taste, and usually is much lighter in color than the other products. Molasses gives feed a sticky touch and appearance and causes grains and other ingredients to clump together, producing a lumpy rather than a free-flowing feed.

Molasses can be dried on an absorbent ingredient such as soy hulls or corn cobs to produce a free-flowing molasses. This produces a light to dark brown meal characterized by a distinct molasses aroma. Heating water-soaked feed containing molasses enhances the aroma.

Stereoscopic Features

If added to a feed in significant quantity the sticky molasses coating on grains and other particles is quite evident. Frequently a particle of grain coated with molasses will be covered by a layer of adhering fine particles and be completely unrecognizable. Also common are balls or masses of the fines and small fragments held together by molasses, the sticky center becoming evident when pressed with forceps.

Particles of dried molasses break up slowly in water, with the mass appearing to melt down and spread out. Color from the mass diffuses into the surrounding area rapidly.

Histological Features

Dried molasses particles appear as brown granular masses with the color spreading slowly into the surrounding medium. Some sugar granules may be found in and around the dispersing masses.

Pomaces

Pomaces are dried mixtures of skins, pulp, and crushed seeds. They may be used fresh, ensiled, or dried. Pomaces may be produced from tomatoes, apples, grapes, and other fleshy fruits and vegetables.

Definitions

Dried apple pomace is the sound, dried residue obtained by the removal of cider from apples. Rice hulls may be a common adulterant.

Dried tomato pomace is the dried mixture of tomato skins, pulp, and crushed seeds.

Dried grape pomace is the refuse from grape juice and wine production, consisting primarily of skins, seeds, and stems.

Macroscopic Features

Tomato pomace meal is reddish colored with flat seeds and smooth, lustrous skin fragments. The seeds are covered with a dense coat of hair-like rods. Apple pomace meal is flaky and brown, with a distinctive cider odor. Some flakes resemble wheat bran particles. Grape pomace meal is dark brown with scattered pieces of stem among the seeds, skins, and pulp.

Stereoscopic Features

Tomato pomace contains soft shiny skins usually rolled into crescents. The seeds are round and flat with a coating of tomato pulp; and stems are pale, fibrous sticks within the meal. Apple pomace is a consistently dusty flake with some darker areas. Grape pomace contains crushed seeds with small, porous holes in their surface. Some of the seeds will still have small linters attached. (Figs. 2.86, 2.87, and 2.88).

Fig. 2.86. Tomato pomace. © Her Majesty the Queen in Right of Canada (2010).

Fig. 2.87. Grape pomace. © Her Majesty the Queen in Right of Canada (2010).

Fig. 2.88. Apple pomace. © Her Majesty the Queen in Right of Canada (2010).

Histological Features

The skin in tomato pomace has no distinguishing cell structure. The cells in the outer epidermal layer of the seed have thick, sclerenchymized walls; hair-like rods support the radial walls of these cells.

Screenings

Screenings are the by-product obtained from the cleaning of grains and other agricultural seeds. Screenings may include light (weight) and broken grains and agricultural seeds, weed seeds, hulls, chaff, joints, straw, elevator or mill dust, sand, insects, and dirt.

No grade of screenings may contain any seeds or other material in amounts that would be either injurious to animals or that will give their milk or flesh objectionable odor or taste. Screenings cannot contain more than four whole prohibited noxious weed seeds per pound nor more than 100 whole restricted noxious weed seeds per pound. The prohibited and restricted noxious weed seeds are determined by the state, province, or country where the screenings are sold or used.

Screenings are typically blended back into the parent by-product after milling or processing.

Definitions

Grain screenings must consist of 70% or more grains, including light and broken grains, wild buckwheat, and wild oats. The ash content must not exceed AAFCO definition.

Mixed screenings is screenings excluded from the preceding definition. The ash content must not exceed AAFCO definition.

Chaff and/or dust is material that is separated from grains or seeds in the usual commercial cleaning processes. It may include hulls, joints, straw, mill or elevator dust, sweepings, sand, dirt, grains, and seeds. If it contains more than 15% ash, the words "sand" and "dust" must appear on the label.

Macroscopic Features

The category or definition into which a particular batch of screenings will fall depends on the amount of the predominant grain, seed, and inert materials present. These products generally contain sufficient amounts of grain, seeds, or hull fragments for identification (Fig. 2.89).

Fig. 2.89. Screenings. © Her Majesty the Queen in Right of Canada (2010).

Stereoscopic Features

Screenings can be identified by several characteristic features: weed seeds in by-products, immature, shriveled and broken grains or seeds of the crop from which the screenings were separated; and an abundance of hulls or glumes, broken awns, rachillae, epithelial hairs; and the absence of most of the grain from the parent crop.

Histological Features

Because screenings are composed of grains and seeds described elsewhere in this manual, refer to the individual grains for histological features.

Triticale

Triticale is a synthetic hybrid species derived by crossing either durum or bread wheat and rye. It is identifiable generally as a whole seed; it exhibits the dorsal shriveling of rye and the color of wheat. The seed is often poorly filled or shrunken but may be larger or intermediate in size compared to the parent species.

3 Feed Ingredients of Animal Origin

Introduction

This chapter provides descriptions of a number of feed ingredients of animal origin and is complementary to Chapter 5, Detecting Animal Products in Feeds and Feed Ingredients. This chapter provides detailed descriptions of individual ingredients whereas Chapter 5 focuses on the methodology used to detect minute quantities (e.g., contamination) of animal products in finished feeds and feed ingredients.

Mammalian Meat Meal and Meat and Bone Meal

Animal waste products from the food industry are often sent to rendering facilities to produce high protein feed ingredients. The rendering process separates fat from other tissues and reduces the moisture content to a low level to prevent spoilage. Fat is then separated to produce various products, such as tallow, and the remaining meat and bones are ground to produce meat meals and meat and bone meals. Continuous dry rendering is the most widely used process.

The quality of meat meal and meat and bone meal is determined not only by the processing, but also by the amounts and types of contaminants and adulterants present. A wide variety of contaminants and adulterants can be found; the microscopist should be familiar with the definitions as well as other potential extraneous materials.

Meat meal and meat and bone meal may be confused with closely related feed ingredients such as tankage, poultry by-products, and dark colored vegetable by-product meals.

Definitions

Mammalian meat and bone meal (often referred to simply as Meat and Bone Meal) is rendered product from mammal tissues, including bone, exclusive of blood, hair, hoof, horn, hide trimmings, manure, stomach, and rumen contents except in such amounts as may occur unavoidably in good processing practices. It shall not contain added extraneous materials not provided for in this definition.

Mammalian meat meal (often referred to simply as Meat Meal) is the rendered product from mammal tissues, exclusive of blood, hair, horn, hoof, hide trimmings, manure, stomach, and rumen contents except in such amounts as may occur unavoidably in good processing practices. It shall not contain added extraneous materials not provided for in this definition.

Macroscopic Features

Meat meal and meat and bone meal are buff to dark brown granules ranging from powder to approximately 8 mesh. Their color, texture, and uniformity vary widely. More bone particles are seen in meat and bone meal than in meat meal alone. Both products are slightly greasy due to residual fat. The odor is that of cooked meat and fat but must not be rancid.

Stereoscopic Features

Larger particles are rough and slightly greasy, with fines clinging to them (Fig. 3.1).

Bone fragments are chunky white, grey, or light tan rock-like pieces. Mammalian bones are dense and break into chunks (Fig. 3.2); less dense poultry bones break with more splinters and with sharper angles. Mammalian bone fragments are very difficult to break using forceps. Poultry bone in meat meal does not necessarily indicate the presence of poultry by-product meal, but may indicate the added bone from chicken deboning operations.

The higher rendering temperatures used in Europe are sufficient to hydrolyse the hair fragments making them unrecognizable as to species under the microscope. Hair is almost always present in meat meals and meat and bone meals produced in North America (Fig. 3.1). Some hair is hydrolyzed and added back to meat and bone meals. These hairs fragment easily when ground (or pressed with a probe) and resemble short cylinders.

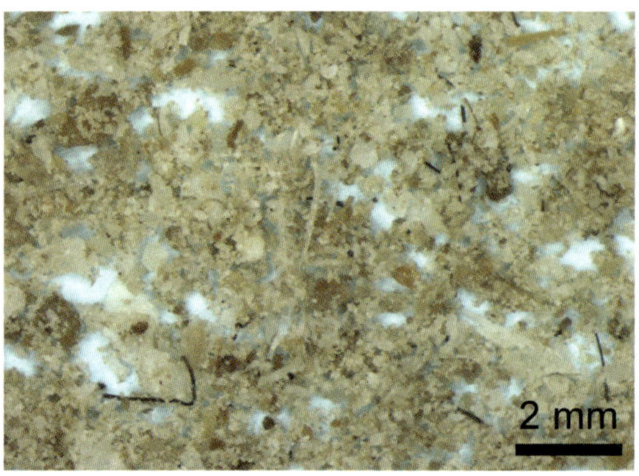

Fig. 3.1. Bovine meat and bone meal observed using a stereoscope. The white particles are the bone fragments whereas the tan colored fragments are muscle and other soft tissues. Numerous hair fragments ranging from white to caramel color to black are also present. © Her Majesty the Queen in Right of Canada (2010).

Histological Features

Although both smooth and striated muscle (Fig. 3.3) occur in meat and meat and bone meal, they are not unique characteristics of these ingredients. Muscle is found in many feed ingredients of animal origin, such as poultry meal and fish meal. Bone, hoof, horn, blood, and hair are more useful characteristics when identifying mammalian meat meal and meat and bone meal.

Bone is composed of a series of extremely thin plates or bars (lamellae) that are arranged in rings around small channels where blood vessels run in the direction of the long axis of the bone (Harvesian canals).

Fig. 3.2. Bone fragments isolated from a mammalian meat and bone meal observed using a stereoscope. © Her Majesty the Queen in Right of Canada (2010).

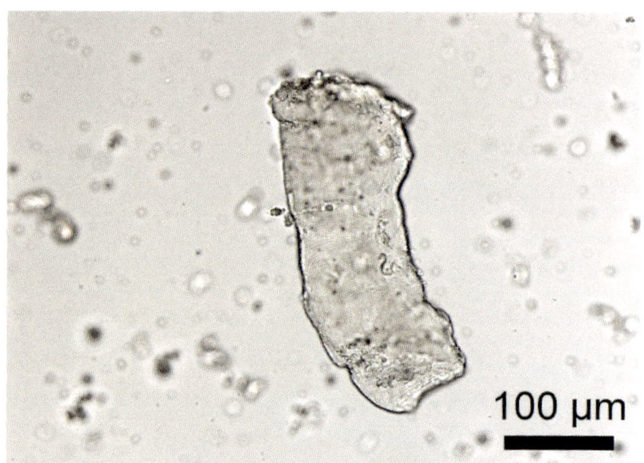

Fig. 3.3. A striated muscle fragment isolated from a bovine meat and bone meal observed using a compound microscope. © Her Majesty the Queen in Right of Canada (2010).

Between these plates are air spaces (lacunae), which are connected to each other by numerous small canals (canaliculi). Nutrients flow through these canals to all parts of the bone. At 100× using brightfield light, the lacunae appear black against a grey background, becoming the most prominent feature of bone (Figs. 3.4, 3.5).

Mammalian bone fragments tend to fragment as blunt, chunky pieces with few sharp edges. The lacunae are generally elliptical to almost globular in shape (width to height ratio from 1.2 to 2.5). The number (i.e. density) of the lacunae of a bone fragment varies from very few to numerous according to the type of bone (spongy or compact). The pattern of distribution is irregular for spongy bone fragments and more organized in the compact bone area from the long bones. The network of canaliculi is fine and can be difficult to detect, but are usually present and visible. Special note of the lacunae located close to Haversian canals: some of them may appear as very elongated (ratio >2.5) in one direction with canaliculi radiating at a right angle making them resembling fish lacunae. This is particularly true when they are observed on very small bone fragments that do not allow detecting the circular distribution pattern around the Haversian canal.

Mammalian bone fragments are difficult to distinguish from avian bone fragments, particularly when only a few bone fragments are available for observation. It is recommended that bone fragments be identified as "terrestrial" if the distinction between mammalian and avian cannot be determined. Fish bone fragments are much more easily distinguished from mammalian and avian bone.

Animal hairs appear as long, thick rods or tubes, depending upon the visibility of the medulla (Figs. 3.6 to 3.10). They are

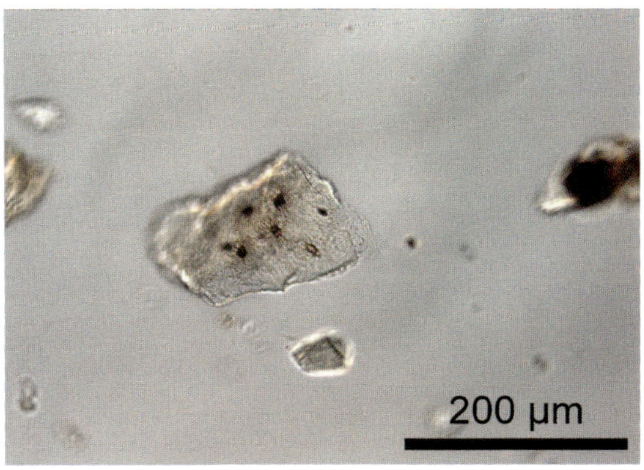

Fig. 3.4. Mammalian bone fragment observed using a compound microscope. The lacunae are visible as black circles and the lacunae are not detectable. © Her Majesty the Queen in Right of Canada (2010).

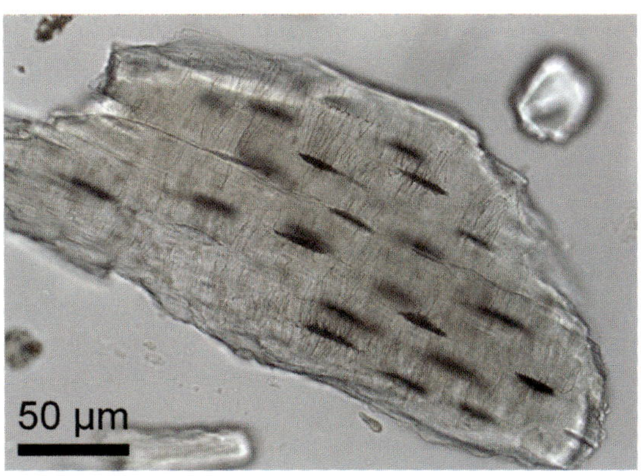

Fig. 3.5. Mammalian bone fragment observed using a compound microscope. The canaliculi are visible as thin lines radiating at a right angle from the elongated lacunae. © Her Majesty the Queen in Right of Canada (2010).

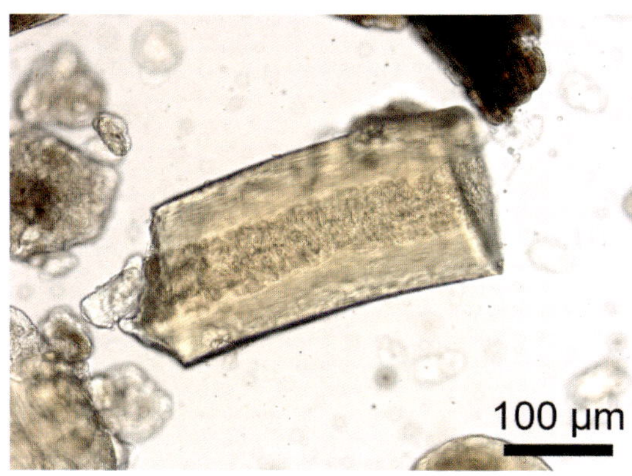

Fig. 3.6. Porcine hair fragment observed using a compound microscope. The medulla and the cortex of the hair fragment are clearly visible even though this hair fragment was subjected to the rendering process. © Her Majesty the Queen in Right of Canada (2010).

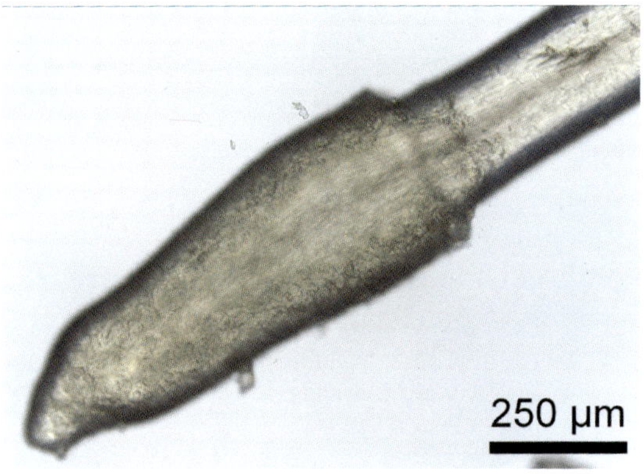

Fig. 3.7. Root end of a porcine hair fragment with follicular material attached observed using a compound microscope. © Her Majesty the Queen in Right of Canada (2010).

slightly curved to severely bent, recognizable by the cuticle of overlapping scales. If the hair is entire, the enlarged root end with follicular material attached will be included (Fig. 3.7).

Cotton fiber should not be confused with hair or wool since it is a flattened fiber with an obvious twist. Cotton fibers birefringe blue and orange when observed with the compound microscope using polarized light (Figs. 3.11, 3.12).

More details regarding hair characteristics are found in Chapter 5.

Hoof or horn may appear as dark grey or black particles with numerous parallel lines on the surface, or they may be distinguished

Fig. 3.8. Scale pattern of an equine hair fragment observed using a compound microscope. © Her Majesty the Queen in Right of Canada (2010).

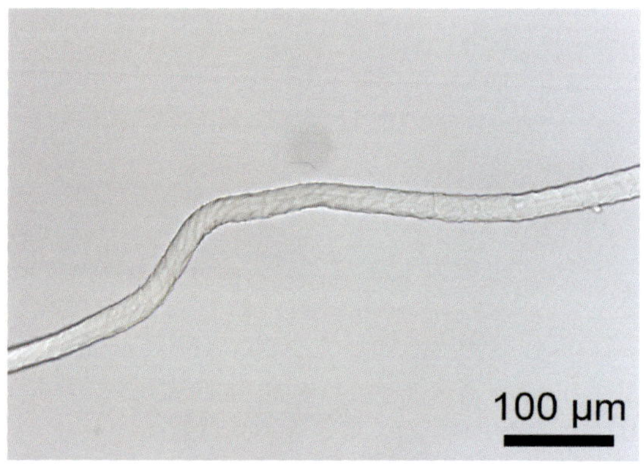

Fig. 3.10. Ovine hair fragment observed using a compound microscope. Note the bends in the hair fragment and the lack of medulla. © Her Majesty the Queen in Right of Canada (2010).

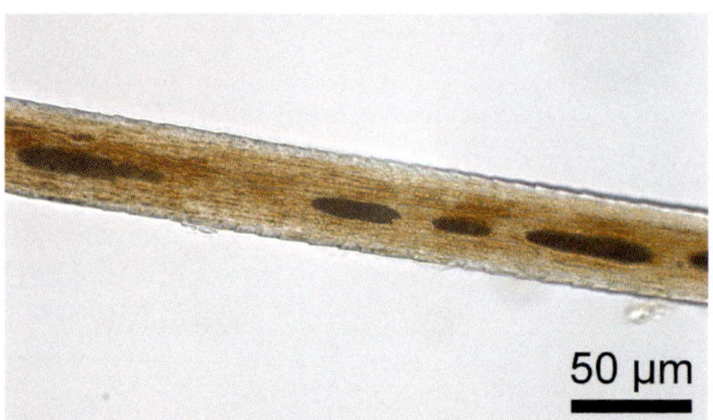

Fig. 3.9. Bovine hair fragment observed using a compound microscope. The fragmented medulla is visible along with the pigmentation in the cortex of the hair. © Her Majesty the Queen in Right of Canada (2010).

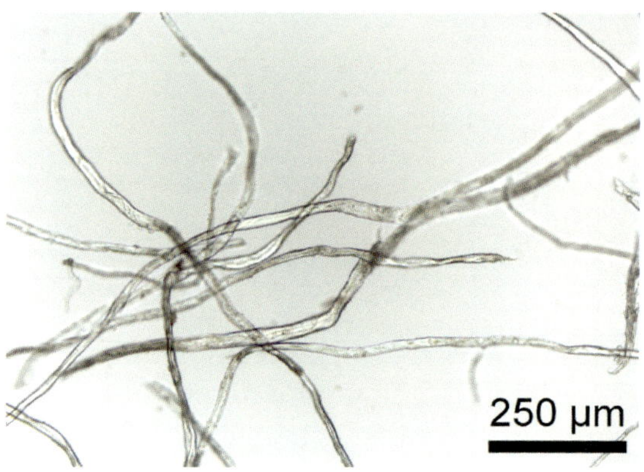

Fig. 3.11. Cotton fibers observed using a compound microscope. © Her Majesty the Queen in Right of Canada (2010).

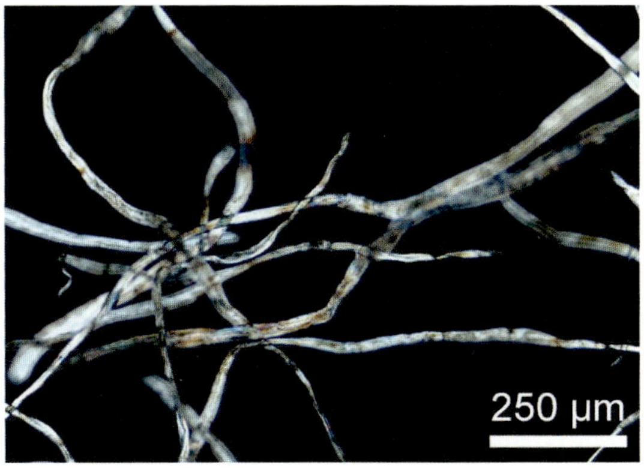

Fig. 3.12. Cotton fibers observed using a compound microscope with polarized light. © Her Majesty the Queen in Right of Canada (2010).

by the wavy, jumbled cleavage lines on broken surfaces. See Chapter 9, Microchemical Spot Tests for a test used to differentiate cartilage from hoof and horn.

Poultry Products

Poultry products are more diverse than other animal products. They include meat, bone, feather, and egg by-products.

Definitions

Poultry by-product meal consists of the ground, rendered clean parts of the carcass of slaughtered poultry such as necks, feet, undeveloped eggs, and intestines, exclusive of feathers, except in such amounts as might occur unavoidably in good processing practices.

Hydrolyzed poultry feathers is the product resulting from the treatment under pressure of clean, undecomposed feathers from slaughtered poultry, free from additives and/or accelerators.

Poultry hatchery by-products is a mixture of egg shells, infertile and unhatched eggs, and culled chicks which have been cooked, dried, and ground with or without removal of part of the fat.

Egg shell meal is a mixture of egg shells, shell membranes, and egg contents obtained by drying the residue from an egg breaking plant in a dehydrator to an end product temperature of 82C.

Macroscopic Features

Poultry by-product meal is lighter in texture, and flakier than meat and bone meal, although color may be similar (Fig. 3.13). The odor is distinct.

Hydrolyzed feather meal is light and fluffy, and the color depends on the source of the feathers.

Hatchery by-product has a salt-and-pepper appearance due to ground egg shells.

Egg shell meal appears as a fine flake.

Stereoscopic Features

Poultry products are identified by finding one or more of the following: feathers, bones, eggs shells, scales, beaks, and claws.

Some feathers resemble hair; others resemble clear plastic tubes. Frequently, the rachis of contour feathers is seen with serrated edges where barbs have broken off. Raw feathers indicate poultry by-product or hatchery by-product meals (Fig. 3.14). Hydrolyzed feathers, coming from a completely different process, indicate blending when found in poultry by-product meals.

Complete hydrolysis changes normal feather structure (Fig. 3.15). Incorrectly processed meals will contain some raw feathers.

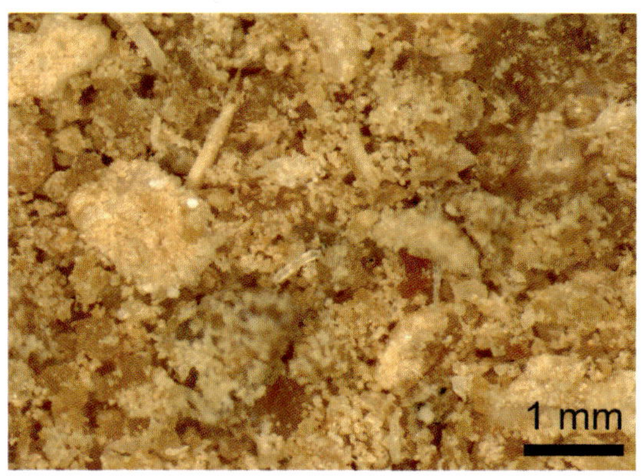

Fig. 3.13. Poultry by-product meal observed using a stereoscope. © Her Majesty the Queen in Right of Canada (2010).

Fig. 3.14. Poultry by-product meal observed using a stereoscope. Raw feather fragments are clearly visible. © Her Majesty the Queen in Right of Canada (2010).

Fig. 3.15. Hydrolyzed feather meal observed using a stereoscope. © Her Majesty the Queen in Right of Canada (2010).

Feather shafts, because of minimal structural changes, are still identifiable after hydrolysis (Fig. 3.16). After processing, the hydrolyzed feathers are ground. The particles appear glassy, tan to brown, or sometimes grey or black, depending on particle size and feather source.

Poultry bones are less dense than mammalian bones and break with sharper angles and splinters (Fig. 3.17). The fragments are more easily broken with a forceps than mammalian bones, but not as easily as fish bones. The bone fragments generally have a light yellow coloring, but can range from white to yellow/tan to light grey.

Scales from the legs and feet are usually seen as chalky white particles, irregularly rounded, and slightly curved at the edges. Particles from the beak and claw are similar to fingernails and are usually present.

Hatchery by-product is differentiated from the other products by the presence of ground egg shell, recognizable by its brittle, calciferous structure and pitted surface.

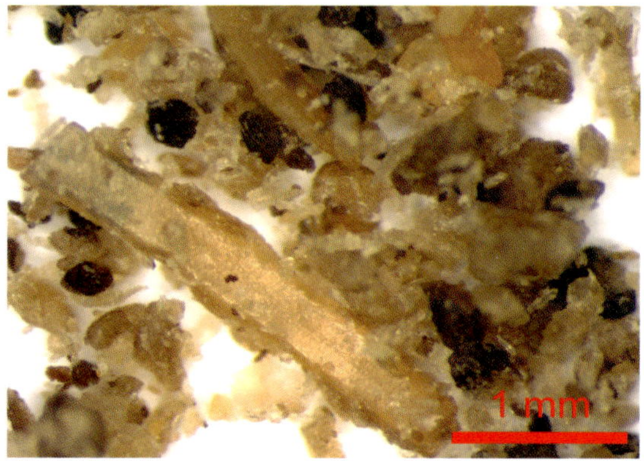

Fig. 3.16. Hydrolyzed feather meal observed using a stereoscope. A feather shaft is clearly visible as the large (c. 2 mm), glassy, tan, cylindrical structure. © Her Majesty the Queen in Right of Canada (2010).

Fig. 3.17. Bone fragments isolated from a poultry meat and bone meal observed using a stereoscope. © Her Majesty the Queen in Right of Canada (2010).

Histological Features

Feather barbs (individual fibers of the vane) are composed of several barbules (Fig. 3.18). The barbules may have tiny hooklets at the nodes. Hydrolyzed feathers generally break at these nodes when ground (Fig. 3.19). At 100×, the hydrolyzed feather mass resembles cooked spaghetti (Fig. 3.20). The individual shaft particles may be long, tubular, oval, flat, or rectangular, but all have vermiform markings on the surface. Individual barbules appear as twisted tubes or rods. Parts of hydrolyzed barbs resemble the individual barbules, but with nodes along them (Fig. 3.20).

Poultry bone is less dense than mammalian bone, and contains larger central canal (medullary cavity) and air spaces. Therefore, the bone walls are thinner and splinter more easily and avian bone fragments tend to have sharp edges. The lacunae range from flattened to oblong (Fig. 3.21) or almost perfectly round in shape (Fig. 3.22). The density of the lacunae generally ranges from several to quite numerous (in general this density is higher than that of mammalian bones). No pattern of distribution can be assigned. The canaliculi of flattened to oblong shaped lacunae are generally present but faintly visible. The canaliculi of globular lacunae may not always be present; when present they tend to be relatively short in length and thicker as compared to those found in mammalian bone fragments.

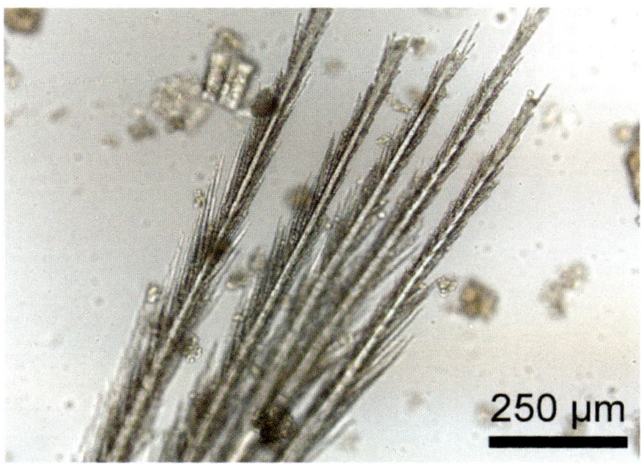

Fig. 3.18. Raw feather fragment observed using a compound microscope. The feather barbs are the primary structures in this image with the barbules attached. © Her Majesty the Queen in Right of Canada (2010).

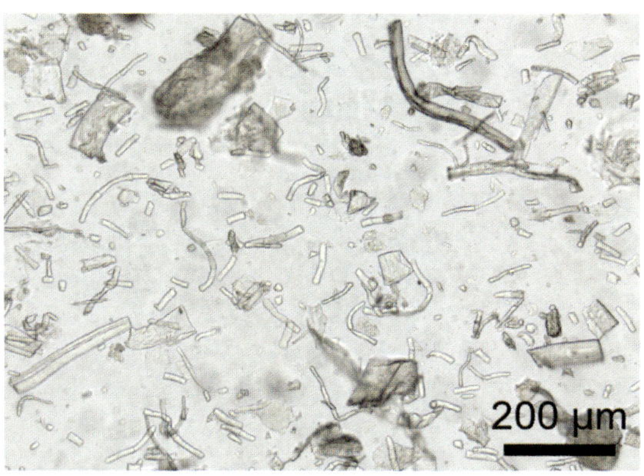

Fig. 3.19. Hydrolyzed feather meal observed using a compound microscope. Most of the feather barbules are broken and only contain a small number of nodes. © Her Majesty the Queen in Right of Canada (2010).

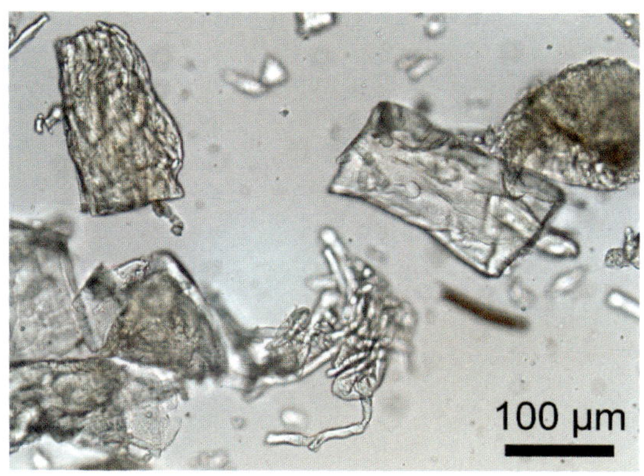

Fig. 3.20. Hydrolyzed feather meal observed using a compound microscope. The rectangular structures at the top of this image are the feather barbs with barbules appearing melted around them. A mass of intertwined barbules at the bottom centre of the image resembles cooked spaghetti. © Her Majesty the Queen in Right of Canada (2010).

Fig. 3.21. Poultry bone fragment observed using a compound microscope. The lacunae of this fragment are flattened to oblong in shape. © Her Majesty the Queen in Right of Canada (2010).

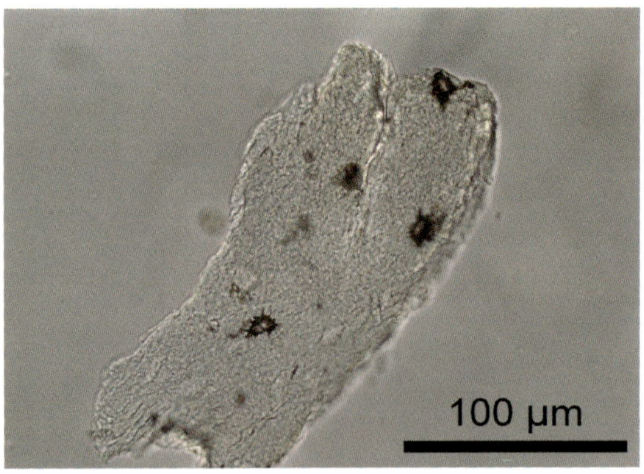

Fig. 3.22. Poultry bone fragment observed using a compound microscope. The lacunae of this fragment are globular in shape. © Her Majesty the Queen in Right of Canada (2010).

Bone Meal

Bone is thoroughly cooked to extract protein and fat, then dried and ground. This product is usually free from heavy contamination, although small quantities of vegetable fiber and hoof can be present.

Definitions

Bone meal, steamed is the dried and ground product sterilized by cooking undecomposed bones with steam under pressure. Grease, gelatin, and meat fiber may or may not be removed.

Bone meal, cooked is the dried and ground sterilized product resulting from wet cooking of undecomposed bones without steam pressure. Fat, gelatin, and meat fiber may or may not be removed.

Macroscopic Features

Bone meal is a dusty white solid ranging in size from fine powder to approximately 8 mesh. It has a unique odor. It is difficult to confuse this product with other organic ingredients; however, it could be mistaken for some inorganic ingredients (e.g., dicalcium phosphate or pulverized limestone).

Stereoscopic Features

The size of the bone particles will be dependent on how finely they are ground. Some bone meals are coarser than others (Figs. 3.23, 3.24). Large particles are white to grey with smooth to moderately rough surface. The fines are a white powder. Bone meal is readily differentiated from limestone and dicalcium phosphate at 10×. Refer to meat and bone meal section for further details.

Histological Features

Refer to meat and bone meal section for details of bone structure.

Fig. 3.23. Coarse ground bone meal observed using a stereoscope. © Her Majesty the Queen in Right of Canada (2010).

Fig. 3.24. Fine ground bone meal observed using a stereoscope. © Her Majesty the Queen in Right of Canada (2010).

Blood Meal

Blood is dried by one of three methods; drum, spray, or ring/flash. The spray and ring/flash drying processes avoid overheating and yield a product considerably higher in available lysine than drum-dried blood.

Drum-dried blood can contain bone and small quantities of hoof, hair, vegetable fiber (paunch or intestinal contents), and/or tankage. Contamination is usually caused by inadequate cleanliness. Spray- and ring/flash-dried products are generally free of contaminants during drying but may have various by-products blended in before shipment.

Due to its characteristic shape, texture, and color, blood meal is not easily confused with other feed ingredients. However, blood meal can be confused with similar dark, rounded, wet-rendered tankage and poultry by-product meal particles, but the latter are rougher in texture and are mottled.

Definitions

Blood meal, conventional cooker-dried is produced from clean, fresh animal blood, exclusive of all extraneous material such as hair, stomach belchings, and urine except in such trace as might occur unavoidably in good manufacturing processes. The product usually has a dark blackish color, and is rather insoluble in water.

Blood meal, flash-dried is produced from clean, fresh animal blood, exclusive of all extraneous material listed above. A large portion of the moisture is usually removed by a mechanical dewatering process or by condensing by cooking to a semi-solid state.

Blood protein is produced by quick freezing and/or transporting in a chilled state, clean, fresh, whole, or dewatered animal blood exclusive of all extraneous material listed above.

Spray-dried animal blood is produced from clean, fresh animal blood, exclusive of all extraneous materials listed above. Moisture is removed from the blood by a low temperature evaporator under vacuum until it contains approximately 30% solids. It is then dried by spraying into a draft of warm, dry air which reduces the blood to finely divided particles.

Macroscopic Features

Drum-dried blood is a dark red to black solid, ranging from powder to ½" irregular or broken pieces often exhibiting conchoidal fractures. The larger pieces have fairly smooth surfaces which are glossy when rubbed on paper. They are not easily crushed or flaked.

Spray-dried and ground ring/flash-dried blood are fine powders, dark red to almost black. Ring/flash-dried blood may have larger globules present, but these are less hard and glassy than drum-dried blood.

Stereoscopic Features

Spray-dried blood occurs as relatively uniform, small, spherical particles that are easily crushed (Fig. 3.25). Ring/flash-dried blood is less uniform, and is often ground to a powder. It is a mixture of small round particles and broken particles. Drum-dried blood is coarse, the most crush-resistant of the blood meals, and the fines from grinding are predominantly broken and irregular (Fig. 3.26).

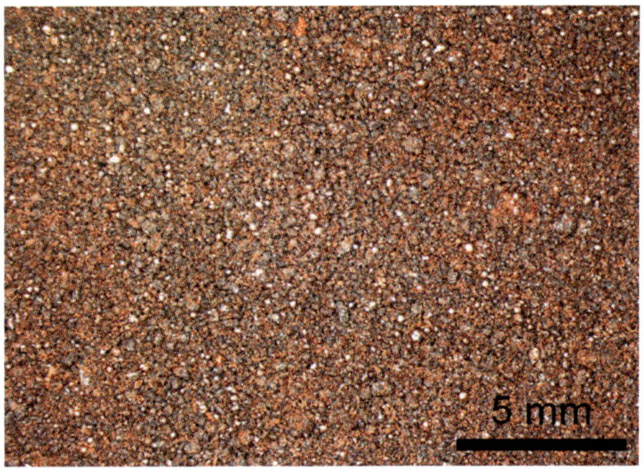

Fig. 3.25. Spray-dried blood meal observed with the stereoscope. © Her Majesty the Queen in Right of Canada (2010).

Fig. 3.26. Drum-dried blood meal observed with the stereoscope. © Her Majesty the Queen in Right of Canada (2010).

Histological Features

Blood appears as roughly rounded, light wine red to ruby red translucent particles which may have a purple tinge at the edges. Tiny black areas on blood meal particles may be due to opacity or charring.

It is recommended that a microchemical spot test be performed to confirm the presence of blood—see Chapter 9, Microchemical Spot Tests, Method XX. If performed using a microscope slide with a coverslip and observed with a compound microscope, the oxygen bubbles formed by the reaction will be captured under the coverslip. The formation of oxygen bubbles along with the blue/green color produced confirms the presence of blood (the use of a positive control is highly recommended). See Fig. 3.27.

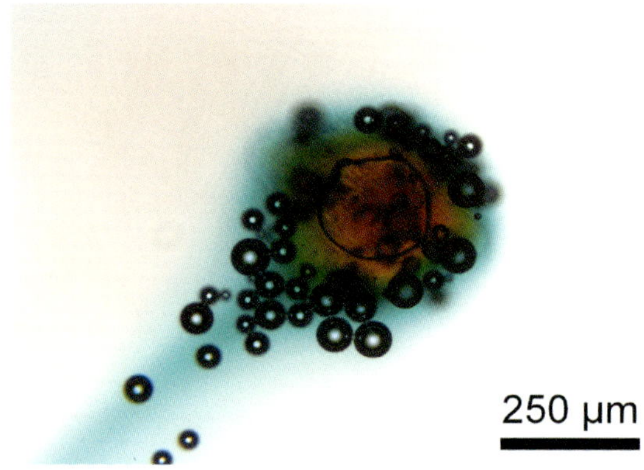

Fig. 3.27. Positive microchemical spot test reaction for suspect blood fragments observed using a compound microscope. The blood particle is red and surrounded by oxygen bubbles. The blue color moving in the direction of the solution is clearly visible. © Her Majesty the Queen in Right of Canada (2010).

Dairy Products

Dried milk products are used extensively in calf and piglet starter feeds but are seldom seen in adult rations or poultry rations.

Definitions

Dried buttermilk, feed grade is the residue obtained by drying buttermilk.

Dried skimmed milk, feed grade is the residue obtained by drying defatted milk.

Dried whey is the product obtained by drying whey.

Dried whey product is the product obtained by drying whey from which a portion of the lactose, protein, and/or minerals have been removed.

Dried whole milk, feed grade is the residue obtained by drying milk.

Macroscopic Features

The various dairy products cannot be differentiated by the feed microscopist. The following are general observations and keys to ingredient identification only. Once a dairy product has been pelleted or extruded in a formula feed, there are no identifiable characteristics remaining.

The dried dairy products used in feeds are soft powders or granules of different degrees of fineness and color. Color may range from white to brown depending upon processing, as overheating can affect the natural color. Roller-processed products have varying amounts of scorched particles and are light cream to light brown in color. Properly processed dried skimmed milk and sweet cream buttermilk are almost white. Acid buttermilk scorches more easily and is yellow. Dried whey ranges from pale yellow through orange to tan, depending upon the dye residues from cheese processing.

Stereoscopic Features

To detect a milk product, sprinkle a small amount of the fine portion of the sample on water and examine against a dark background. Small cloudy streams may be seen flowing from the milk particles.

Histological Features

Milk particles must be observed immediately after a slide has been prepared as they tend to dissolve in mounting media.

Milk particles usually appear yellow to brown with transmitted light and white with reflected light. The particles often appear to have vacuoles. Most milk products contain some fat, indicated by the oil globules present in the slide preparation. Dried whole milk has more fat than dried skimmed milk, and both have more fat than dried buttermilk. Dried whey products contain no oil, but do contain free lactose crystals. These crystals appear as wedges or axe heads, their presence may be enhanced by polarized light (Figs. 3.28, 3.29).

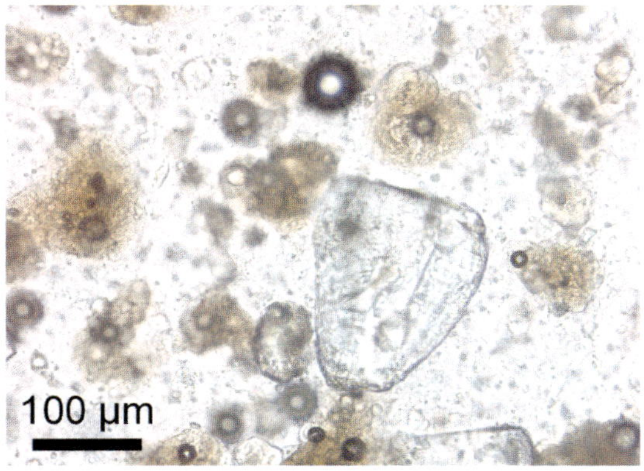

Fig. 3.28. Axe-head shaped lactose crystal observed using the compound microscope with brightfield light. © Her Majesty the Queen in Right of Canada (2010).

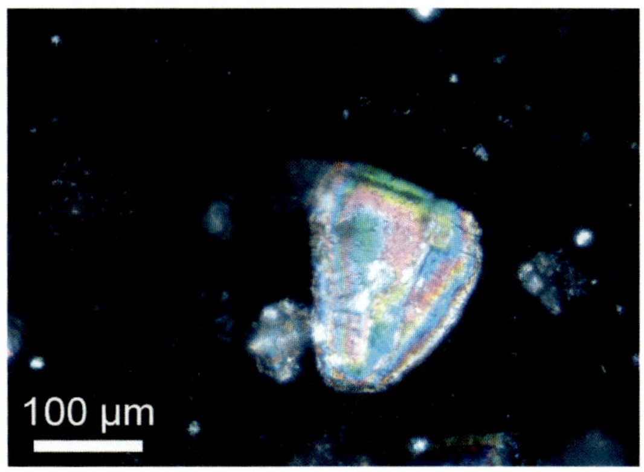

Fig. 3.29. Axe-head shaped lactose crystal observed using the compound microscope with polarized light. © Her Majesty the Queen in Right of Canada (2010).

Meat Meal Tankage and Meat and Bone Meal Tankage

These products are derived from the wet-rendering process, which is not used widely due to economic and environmental concerns. The following information is for reference only.

The same materials used in the production of meat meal and meat and bone meal are partially disintegrated and cooked under pressure. As in meat meal, a wide variety of contaminants can be found in tankage. These include hoof, horn, hair, fluff, scorched meat, and vegetable fiber from paunch and intestinal contents. Small quantities of poultry offal, including feathers and bone, are occasionally present. If the product is made from restaurant waste, traces of glass may be present. Although similar to meat meal, tankage can usually be distinguished by a darker and slightly smoother appearance.

Definitions

Meat meal tankage is the rendered product from mammal tissues exclusive of hair, hoof, horn, hide trimmings, manure, stomach, and rumen contents except in such amounts as may occur unavoidably in good processing practices. It may contain added blood or blood meal, but it shall not contain any other added extraneous materials not provided for in this definition.

Meat and bone meal tankage is the rendered product from mammal tissues, including bone, exclusive of hair, hoof, horn, hide trimmings, manure, stomach and rumen contents except in such amounts as may occur unavoidably in good processing practices. It may contain added blood or blood meal, but it shall not contain any other added extraneous materials not provided for in this definition.

Macroscopic Features

Meat meal and meat and bone meal tankages are medium dark brown to almost black granular solids, ranging from powder to approximately 8 mesh. Most samples are slightly greasy. The odor is stronger than meat meal but should not be rancid or putrid.

Stereoscopic Features

Larger particles are rough and slightly greasy, with fines clinging to them. They have fairly rough surfaces, smoother and rounder than meat meal. Bone fragments are chunky white, grey, or light tan rock-like pieces. Mammalian bones are dense and break into chunks; less dense poultry bones break with more splinters and with sharper angles. Hair is always present.

Histological Features

See histological features of meat meal, meat and bone meal, and blood meal.

Animal Liver Meal

Liver meal is a specialty item with limited use in the pet food industry. It is a relatively clean product; however, bone, blood meal, hydrolyzed feather meal, hair, and vegetable fibers may be found.

Fig. 3.30. Animal liver meal observed with the stereoscope. © Her Majesty the Queen in Right of Canada (2010).

Definition

Animal liver meal is the product obtained by drying and grinding liver from slaughtered mammals.

Macroscopic Features

Liver meal is yellowish brown to dark brown, ranging in size from coarse powdered to approximately 8 mesh. It can be slightly greasy and is similar to meat meal in color.

Stereoscopic Features

Particles are fairly smooth, brown pieces similar to wet-rendered tankage, but lighter in color (Fig. 3.30). They are amorphous, and neither granules nor fines are distinguishable from meat.

Histological Features

Liver is a granular product without distinct features to distinguish it from other meat meals.

4 Feed Ingredients of Marine Origin

Fish Meal and Solubles

Most fish meal is produced by cooking the fish with live steam under pressure and pressing to remove most of the oil and some water. The press liquor (stickwater) may be either evaporated to a solids content of 50% (condensed fish solubles), dried (dried fish solubles), or recombined with the pressed fish in the dryer.

Definitions

Fish meal is clean, dried, ground tissues of undecomposed whole fish or fish cuttings, either or both, with or without part of the oil.

Condensed fish solubles is obtained by condensing the stickwater.

Dried fish solubles is obtained by dehydrating the stickwater.

Fish protein concentrate, feed grade is prepared from clean, undecomposed whole fish or fish cuttings using the solvent extraction process developed for the production of edible whole fish protein concentrate.

Macroscopic Features

Fish meal is light to medium brownish yellow, ranging from a powder to approximately 8 mesh (commercial grind; 2.36 mm). Bones have a pearly luster and are present with scales in considerable quantity—classically bones and scales represent between 10–20% w/w. Most samples are slightly greasy, with an intense fishy odor.

Stereoscopic Features

The most unique components of fish meal are bones and scales. Some bones are slender and pointed, others are vertebral in shape. Almost all show anatomical structure. Scales are flat or curved, glossy translucent structures often with concentric markings (Fig. 4.1). The flesh resembles expeller meat meal, but is lighter in color. The fines are granular. Fragments of gills, teeth, eyes, and otoliths are regularly present and may help, if need be, to in-depth characterization.

Histological Features

Fish meal can be identified by fragments of scale and bone. The bones are generally white to gray, translucent to opaque slivers. Occasionally fish bones may present an amber color

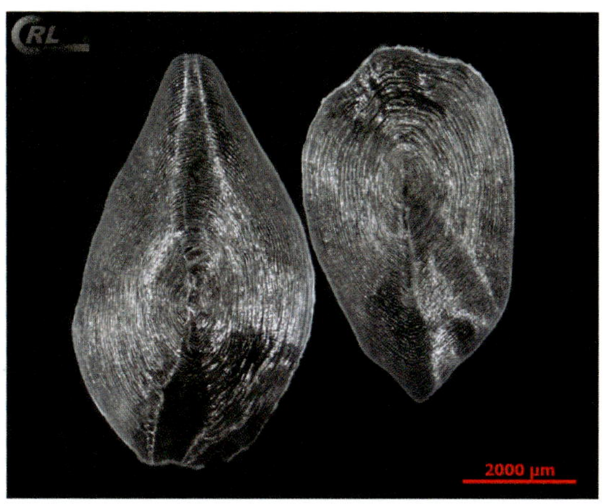

Fig. 4.1. Salmon scales observed using a stereoscope. The concentric annual growth rings can be easily seen. Micrograph by courtesy of CRL-AP.

especially when fish oil has not been thoroughly extracted. The distinguishing feature of bone is the lacuna, which is dark and spindle-shaped with fine wavy lines (canaliculi) radiating from either end (Fig. 4.2). These canaliculi are normally highly visible in fish bone, but not in avian or mammalian bone.

General features of fish bones compared to bones of "terrestrial" animals are described in Chapter 5, Detecting Animal Products in Feeds and Feed Ingredients.

Fish bones' lacunae vary greatly in shape and features. This diversity of histological features is not only observed between taxonomic groups but also between bones from a single species or even inside the same large fishbone fragment. This may be notably explained by the various ossification ways that result in a diversity of skeletal tissues.

Some fish skull fragments are almost free from any lacunae: they are referred to as chondroid bones (intermediate region between cartilage and bone). Fish skull particles present a very lightweight spongy pattern with large holes and conduits, where the circulatory system was located (Fig. 4.3).

Lacunae shape may also be of the "short spindle" type with no visible canaliculi as in tuna or salmon. Some species (e.g., cod) have fish bones without lacunae but present a faint yellow-brownish fibrous look without any other typical markers allowing bone identification (Fig. 4.4). The organization pattern of lacunae, although described as parallel (running all in the same direction of the bone long axis) sometimes can be different. A disorderly crossing of elongated lacunae over each other can be observed on some fish bone fragments

Fig. 4.2. Lacunae and radiating canaliculi in a fishbone observed using a compound microscope. Micrograph by courtesy of CRL-AP.

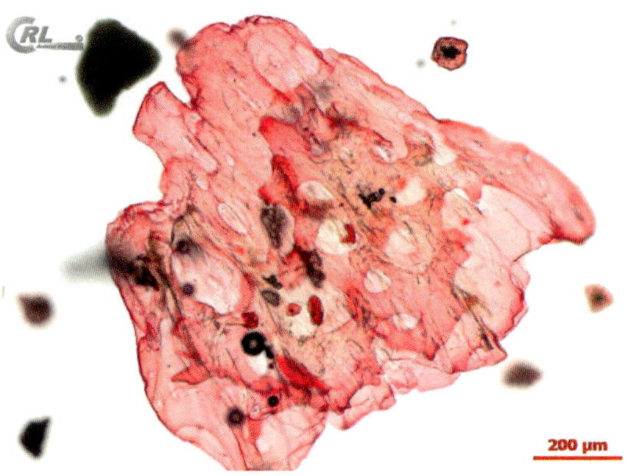

Fig. 4.3. Fish skull bone fragment with large holes. Observed using a compound microscope, stained with Alizarin Red. Micrograph by courtesy of CRL-AP.

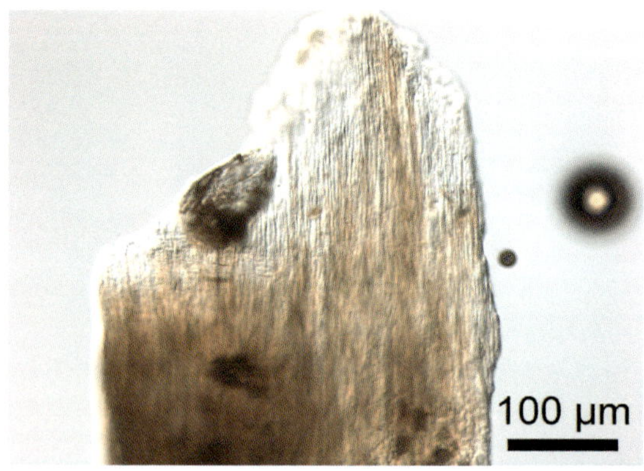

Fig. 4.4. Fish bone fragment with a fibrous appearance observed using a compound microscope using DIC lighting. © Her Majesty the Queen in Right of Canada (2010).

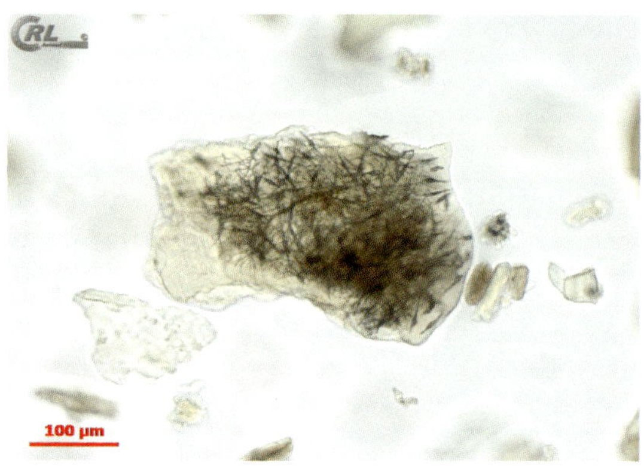

Fig. 4.5. Fish bone fragment with typical disorganized appearance of lacunae observed using a compound microscope. Micrograph by courtesy of CRL-AP.

Fig. 4.6. Gill fragment with distinct three layers observed using a compound microscope, stained with Alizarin Red. Micrograph by courtesy of CRL-AP.

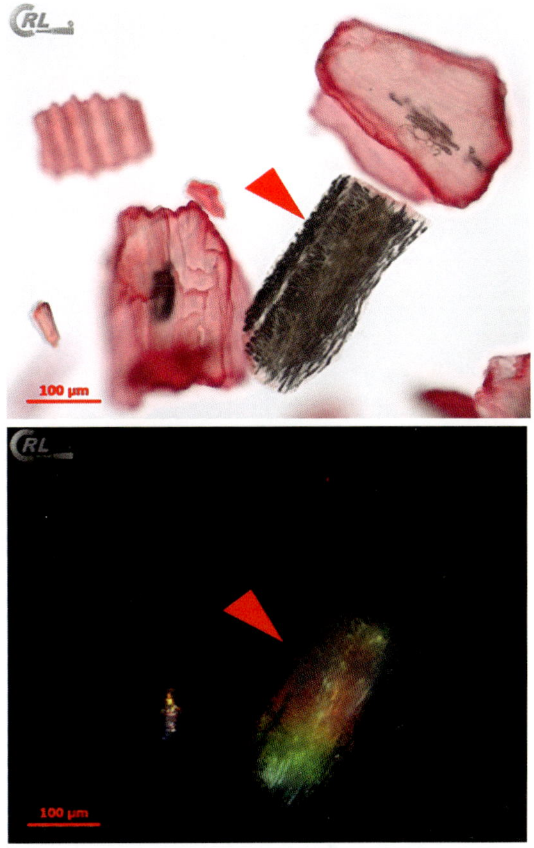

Fig. 4.7. Example of high birefringence of gills. The arrowheads show the same faintly stained gill fragment in bright field and under polarized light. Stained with Alizarin Red. Micrograph by courtesy of CRL-AP.

Fig. 4.8. Otolith observed using a stereoscope. Micrograph by courtesy of CRL-AP.

(e.g., herring) as well as abrupt change of parallel pattern directions (Fig. 4.5). Occasionally, fish bones have well-organized parallel lacunae from which all canaliculi extend to one side only (e.g., herring, pilchard).

Gills are the respiratory organs of fish. They are composed of bony gill arches, cartilaginous primary filaments and soft tissue secondary lamellae, the whole being protected by gill flaps (opercula). Due to their fragile architecture, gills are almost totally destroyed during fish meal processing with the exception of the bone support elements forming the opercula and the arches. Gill fragments appear as structures organized in layers, or palisades, with alignment of large air-filled lacunae. This palisade pattern differs according the taxonomic order to which the fish species belongs. For Cupleiforms (e.g., pilchard, herring, anchovies, etc.), three well-defined layers are observed (Fig. 4.6), whereas for other orders like Salmoniforms (e.g., salmon, trout, etc.) only two less-organized layers are noted. Frequently, the surfaces of gill flap fragments present two different sides: one is flat, the opposite has extruding bone profiles.

Gill fragments are usually birefringent (Fig. 4.7) when observed using polarized light, and, like other bone structures, are frequently stained red with Alizarin Red (Fig. 4.6). Sometimes, when the mineralization of the gill is imperfect, the staining can be very faint or almost absent (Fig. 4.7).

Otolith fragments also can be present in sediment containing fish meals. Otoliths (fish inner ear bones) look like pure white round, oval, or even kidney shaped irregular structures (Fig. 4.8). Their size is very variable depending on the species and the age of the fish and

Fig. 4.9. Otolith fragment observed using a compound microscope. Micrograph by courtesy of CRL-AP.

Fig. 4.10. Otolith fragment from Fig. 4.9 under polarized light. Micrograph by courtesy of CRL-AP.

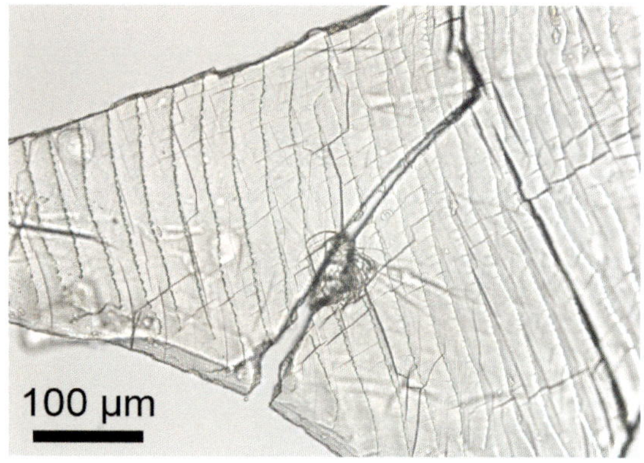

Fig. 4.11. Fish scale observed using a compound microscope. © Her Majesty the Queen in Right of Canada (2010).

Fig. 4.12. Melanophores (pigment dots) within a salmon scale observed using a compound microscope. Micrograph by courtesy of CRL-AP.

ranges from 500 µm up to several centimeters. Otolith fragments observed using a compound microscope appear as wedgelike particles with alternating striping of dark and light bands corresponding to successive deposits of calcium carbonate during their growth (Fig. 4.9). These accrete layers are responsible for a strong structural birefringence when observed under polarized light (Fig. 4.10). Alizarin Red does not stain otoliths.

Scale fragments are normally present with fish bone fragments. They generally appear as flat, translucent fragments, with equidistant, thin parallel lines, which are spaced apart giving a steplike appearance due to the concentric growth rings (Fig. 4.11). Some fragments can appear to have tiny dark pigment speckles lightly peppered throughout the fragment (Fig. 4.12).

Variations in fish body surface exist between bone fish (e.g., herring, salmon, and eels) and cartilaginous fish (e.g., sharks, rays, dogfish).

Bone fish have their surface covered with scales containing at least one layer of lamellar bony materials (therefore, scales always stain with Alizarin Red). Such scales are referred to as cycloid scales (e.g., carp, salmon, trout, herring) with concentric annual growth rings (Fig. 4.1) and smooth margins, or ctenoid scales (e.g., sole, perch) with fine tiny spines on their posterior edge and furrows on the anterior edge. Numerous bone fish species may present both types of scales. The processing of fish meal also results in the reduction of the scales into small fragments, sometimes resembling minute roof tiles attached to each other. Typical examples of such fragments can be found in cod or herring meal (Fig. 4.13).

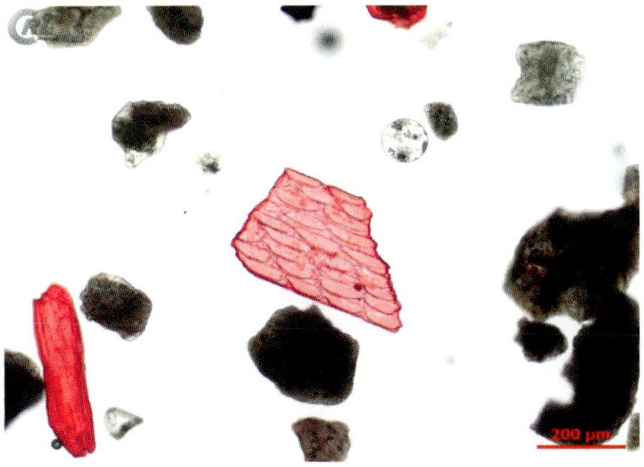

Fig. 4.13. Scale fragment stained with Alizarin Red resembling minute roof tiles attached to each other, observed using a compound microscope. Micrograph by courtesy of CRL-AP.

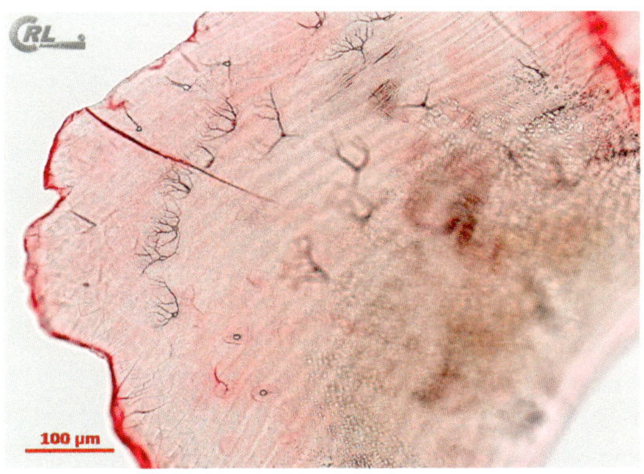

Fig. 4.14. Fish scale presenting lacunae observed using a compound microscope. Stained with Alizarin Red. Micrograph by courtesy of CRL-AP.

Fig. 4.15. Ray skin presenting placoid scales (denticles). Observed using a stereoscope. Micrograph by courtesy of CRL-AP.

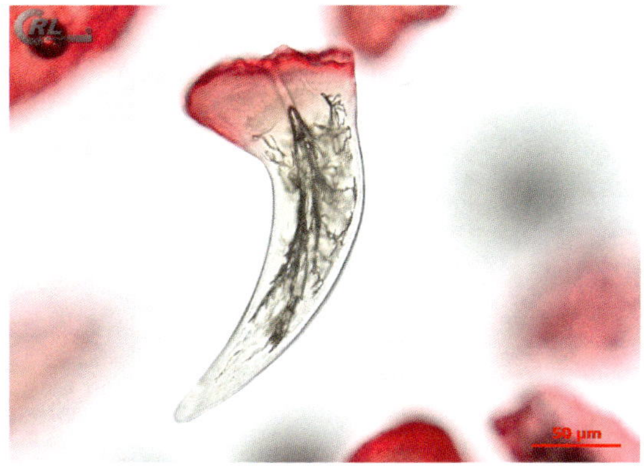

Fig. 4.16. Placoid scale (denticle) stained with Alizarin Red, observed using a compound microscope. Note the typical dendrite structures. Micrograph by courtesy of CRL-AP.

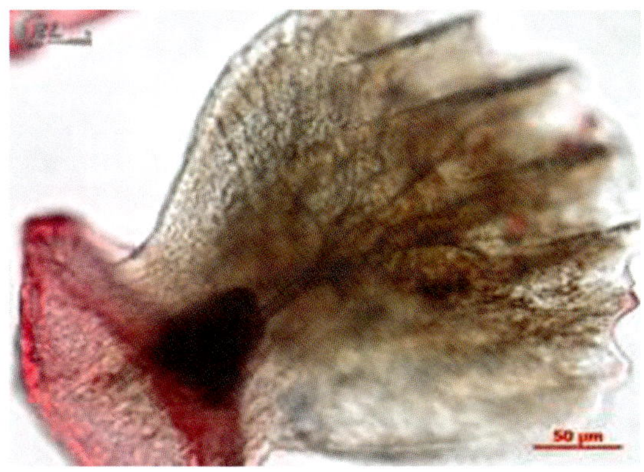

Fig. 4.17. Placoid scale (denticle) stained with Alizarin Red, observed using a compound microscope. Note the typical dendrite structures. Micrograph by courtesy of CRL-AP.

Unusual presence of lacunae in fish scales can be observed (Fig. 4.14).

Cartilaginous fish only have placoid scales (denticles or skin teeth) (Fig. 4.15). These are composed of dentine covered with a thin layer of enamel. Therefore, they appear red with a whitish cover or tip when stained with Alizarin Red. Denticles also often show typical dendrite structures (Figs. 4.16 and 4.17). The shape of denticles is known to be species-specific. Thus, there may be a great diversity of denticles within a fish meal composed of various species being processed.

Teeth can also be detected in fish meals. Teeth are usually found detached from the jawbones of fish. On rare occasion teeth may be still attached (Fig. 4.18).

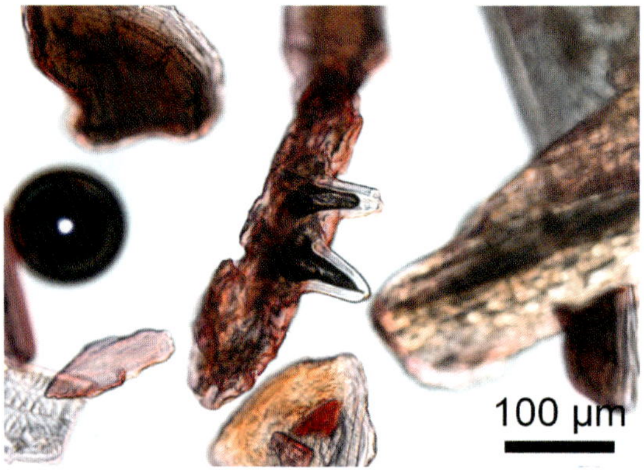

Fig. 4.18. Two teeth attached to a piece of jaw isolated from a fish meal observed using a compound microscope, stained with Alizarin Red. © Her Majesty the Queen in Right of Canada (2010).

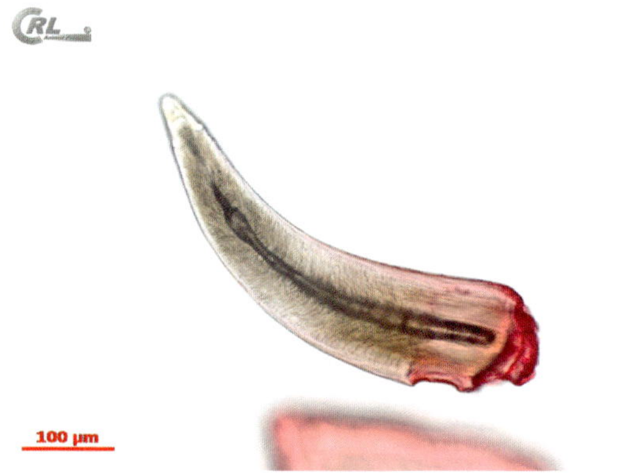

Fig. 4.20. Tooth observed using a compound microscope. This particle was stained with Alizarin Red. The pulp cavity is visible as well as the small enamel tip. Micrograph by courtesy of CRL-AP.

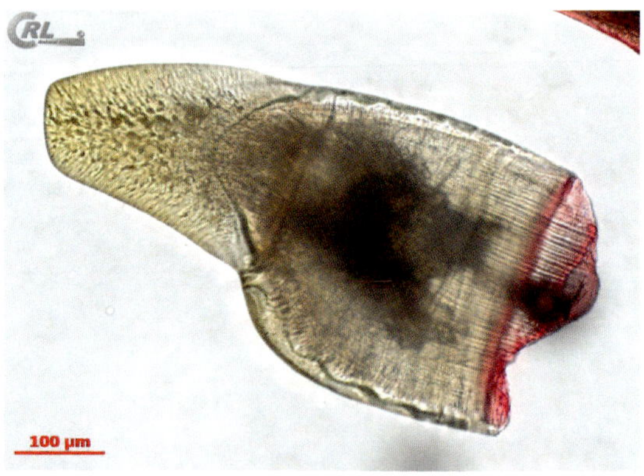

Fig. 4.19. Tooth observed using a compound microscope. This particle was stained with Alizarin Red but it only stains at the base. Note the dentine tubular pattern. Micrograph by courtesy of CRL-AP.

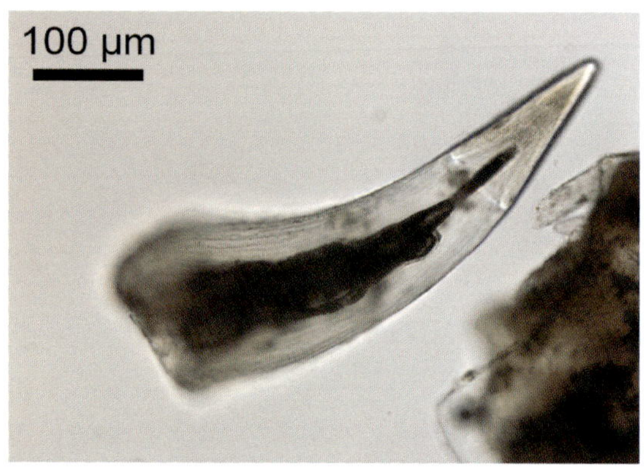

Fig. 4.21. Tooth isolated from a fish meal observed using a compound microscope. © Her Majesty the Queen in Right of Canada (2010).

The structure and composition of a typical fish tooth is very similar to that of a mammalian tooth. The tooth contains a central pulp, a dentine layer, and a cover crown or cap made of enamel (the cap is not always present).

Teeth from fish present a large diversity of shapes (stump, flat, or pointed) and sizes (from several cm or mm to a few 100 µm). (Figs. 4.19–4.21). Nevertheless, the most frequently observed are usually pointed at the tip and have a hollow middle of the tooth where the pulp was found visible (Figs. 4.20 and 4.21). Sometimes, the vascular ducts are visible within large teeth. When present, the enamel cap is clearly separated from the dentine. The tubular aspect of dentine is seldom visible (Fig. 4.19). As with other bony structures, teeth are stained red by Alizarin Red, with the exception of the enamel that remains unstained. Due to the high density and hardness of teeth, coloration is seldom uniform.

Note the confusion that could easily occur between tooth and denticle fragments or spines from ctenoid scales. Although spines from ctenoid scales never present enamel and always have a plain structure (without pulp) that differentiates them from teeth, confusion is common. Such errors are not critical, especially as feed microscopists are required to detect only the presence of fish in a feed matrix.

Lenses of fish eyes can also be detected in fish meals. The lens is an almost perfect hard spherical structure (Fig. 4.22) that can be observed as a whole for small-sized fish species.

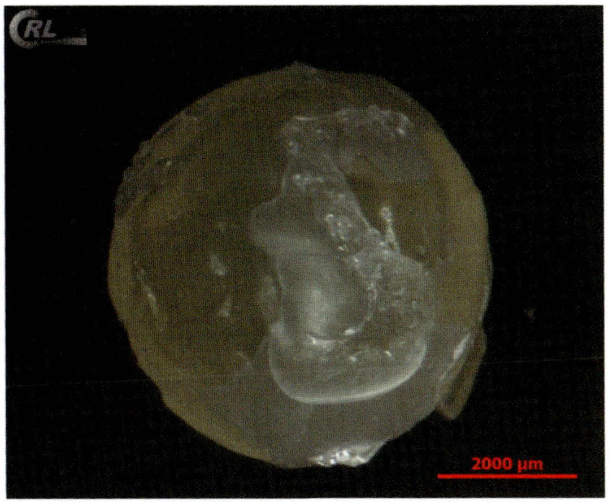

Fig. 4.22. Salmon lens observed using a stereoscope. Micrograph by courtesy of CRL-AP.

For large fish, fragments of lenses can be observed in the sediment. Under the microscope, lens fragments show a multiple layered structure (Fig. 4.23). No obvious birefringence can be noted when illuminated using polarized light.

Crab Meal

Crab meal is produced from the commercial harvesting and processing of crab and related marine organisms. It is used primarily in fish feed.

Definition

Crab meal is the undecomposed ground waste of the crab and contains the shell, viscera, and part or all of the flesh.

Macroscopic Features

Crab meal is a combination of soft tissues (meat) and exoskeleton (shell), with shell predominating. Meal color ranges from tan to light brown, depending on the species harvested. Shell color is generally white, with touches of pigmented spots (orange, dark brown or greenish according to the species). The aroma is distinct.

Stereoscopic Features

Crab meal is recognized by the presence of numerous opaque particles of shell. They are white or cream, with tinges of color. The

Fig. 4.23. Salmon lens observed using a stereoscope. Micrograph by courtesy of CRL-AP.

meat particles are brown, but cannot be distinguished from other animal meals. Some of these particles resemble corn and wheat bran. They curl slightly at the margins, but are not cellulosic. Often crab shell particles present small spiny calcareous outgrowths, which are common with crustaceans.

Histological Features

Crab meal is identified by small, irregular pieces of the chitinous shell. These are extremely variable in thickness, depending on the amount of calcium carbonate deposits within the lamellae. Not all shell fragments are identifiable except the ones originating from feelers, legs, chelae (pincers), or back carapace. These fragments show an outside layer that is usually porous, and resemble honeycomb, round cell patterns. Some sections of this layer are so thin that the pores appear as a single layer of thick-walled cells. These structures are not, or are barely, birefringent when observed using polarized light.

Shrimp and Krill Meals

Shrimp and krill meal is produced from the commercial harvesting and processing of shrimp and related marine organisms. It is used primarily in fish feed.

Definitions

Shrimp meal is the undecomposed ground dried waste of shrimp, and contains parts and/or whole shrimp.

Krill meal is produced from entire freshly caught krill by a drying process (freeze dried). Krill meals mainly originate from South Antarctic fishing campaigns.

Macroscopic Features

Shrimp meal is characterized by numerous particles of the thin mica-like exoskeleton (shell) with the meat. Small black specks (parts of eyes) are common in the meal. Some meals are primarily shells.

Krill meal roughly presents the same macroscopic features as those of shrimp meal except for its color that is usually strongly orange to pink due to its high carotenoid content.

Stereoscopic Features

Shrimp meal is distinguished by fragments of translucent, chitinous shell, and by fragments of segmented structures (e.g., antennae, legs). Numerous spindle-like straight fragments of bristles and tactile appendices are generally present (Fig. 4.24).

Krill meal resembles shrimp meal. The differences that can be used to help distinguish it from shrimp meal are described as: (1) more globular aspects of the particles as they are freeze dried whole animal fragments; and (2) a reduced proportion of translucent shell fragments (Fig. 4.25). Meat fibers are also easily identifiable, and the overall color is much more intense compared to shrimp meals.

Fig. 4.24. Shrimp meal observed using a stereoscope. Micrograph by courtesy of CRL-AP.

Histological Features

Both shrimp and krill meals are characterized by the presence of more or less transparent particles of the chitinous shell (Fig. 4.26). These usually show very fine lines intersecting at random angles and extending across the whole particle. Occasional lines connect across several other lines, forming triangles and other geometric shapes. In some areas there may be cross-hatching (Fig. 4.27). In addition, shell fragments frequently present some rosette markings at their surfaces. This is caused by either cyst-forming parasites or calcium carbonate deposits. Such rosettes can display slightly radial birefringence when observed using polarized light, giving them an appearance like wheel spokes (Fig. 4.28).

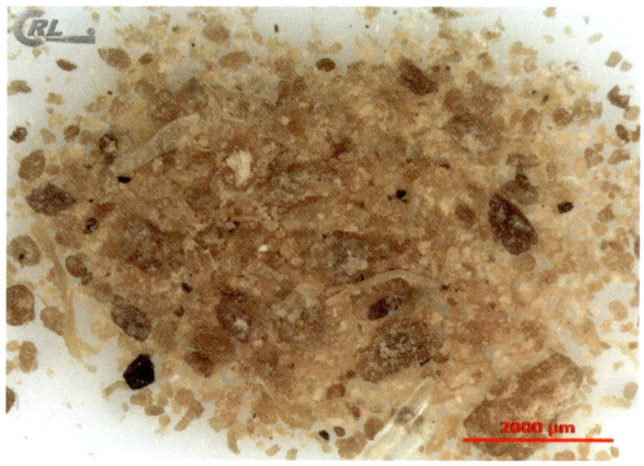

Fig. 4.25. Krill meal observed using a stereoscope. Micrograph by courtesy of CRL-AP.

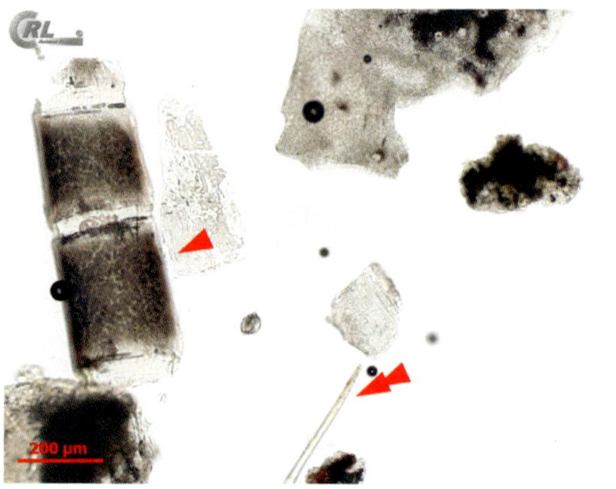

Fig. 4.26. Shrimp meal observed using a compound microscope. Feelers (arrowhead) and straight bristle fragments (double arrowhead) are visible. Micrograph by courtesy of CRL-AP.

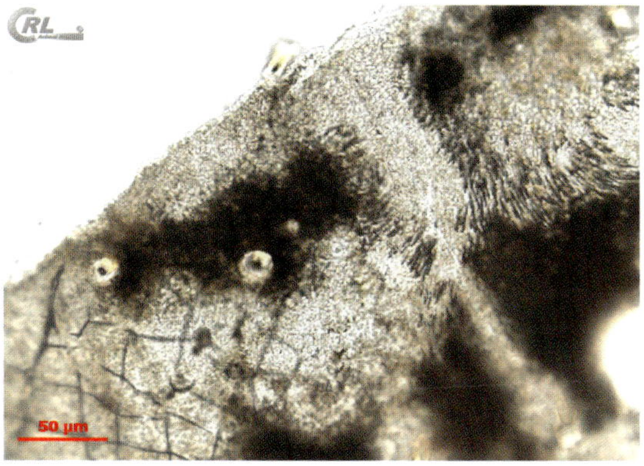

Fig. 4.27. Shrimp shell fragment observed using a compound microscope. The cross-hatched pattern is visible as well as chromatophores and insertion regions of bristles (circular areas with black center). Micrograph by courtesy of CRL-AP.

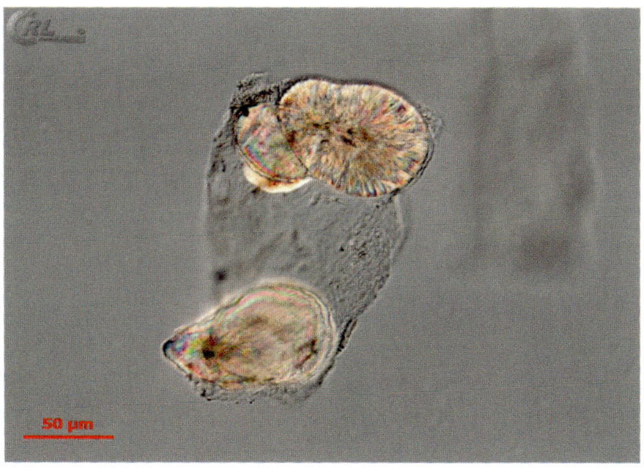

Fig. 4.28. Rosettes, or radial shapes, and polarized deposits from shrimp shell fragments. Observed using a compound microscope with DIC lighting. Micrograph by courtesy of CRL-AP.

The segmented antenna, feathery gill tissue, and heavy rectangular amber cells of the compound eyes are other tissues used to identify shrimp meal.

Marine Invertebrate Meals

Sea invertebrates, mainly mollusk meals, are often used in animal feeds—primarily in fish feeds, but also for poultry feed for layers. Different products are processed and available. Among the most frequently processed are mussel meals, oyster powders, and squid meals.

Definitions

There are two types of mussel meals. ***Mussel meal*** *in se* is processed from the edible part of the bivalves and hence contains meat and very few shell fragments remain. Mussel meal is a high protein and calcium containing feed ingredient sometimes used for partial replacement of fish meals.

Mussel shell powder is a second type of produced feed compound from the shell fraction only. It is used for its high calcium content for layers.

Oyster powder consists of powdered oyster shell and is mainly used as a source of calcium.

Squid meal is prepared from molluscs caught in the open sea and processed into meal. Squid meal is used for its very high crude protein content in substitute of fish meal in aquaculture. Squid oil is also produced.

Macroscopic Features

Mussel meal is a fairly amorphous pale green-yellow to orange powder with a typical shellfish odor.

Squid meals are coarse powders from orange to red in color with a strong fish odor. Unlike fish meals, squid meals do not have any fish bones or scales. Some glossy hard fragments resembling cut nails are visible.

Mussel shell and oyster powders are fine powders of a greyish-white color. Depending on the grinding process, shell fragments with a typical glossy aspect are sometimes identifiable.

Stereoscopic Features

Mussel meals, mussel shell powders, squid meals, and oyster powders are all lacking well-characterized features allowing identification with a stereomicroscope. Nevertheless, the presence of iridescent flat particles of nacre (mother-of-pearl) sometimes enables the determination of the shell origin.

Histological Features

Mussel meals always contain varying amounts of shells. Therefore, the presence of mussel

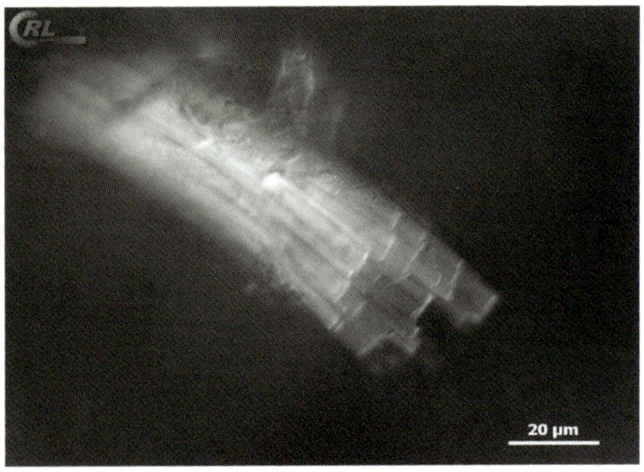

Fig. 4.29. Mussel nacre particle, detail of prismatic elongated fragments of nacre. Observed using a compound microscope with DIC lighting (black and white picture). Micrograph by courtesy of CRL-AP.

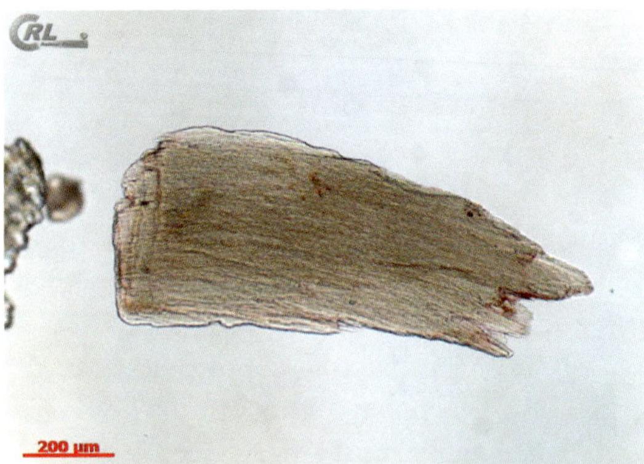

Fig. 4.30. Oyster shell fragment observed using a compound microscope. Micrograph by courtesy of CRL-AP.

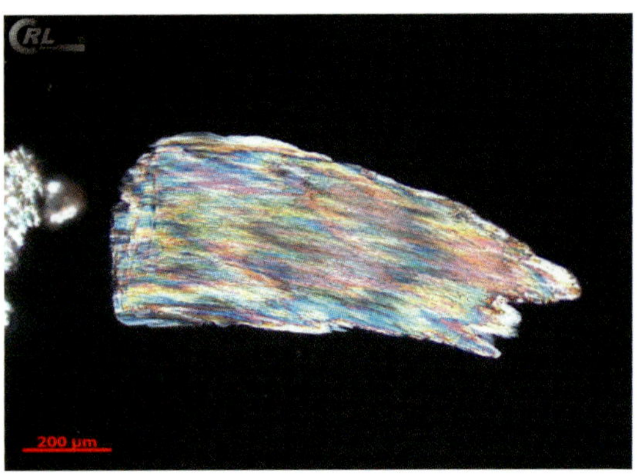

Fig. 4.31. The same oyster shell fragment as in Fig. 4.30 observed using a compound microscope and polarized lighting. Micrograph by courtesy of CRL-AP.

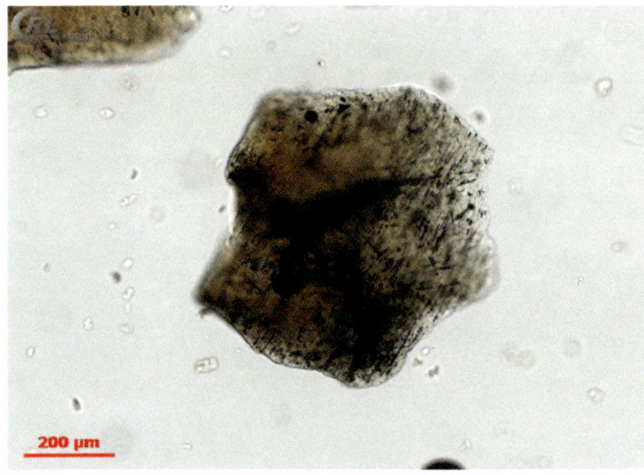

Fig. 4.32. Typical fragment from squid meal observed using a compound microscope. Micrograph by courtesy of CRL-AP.

meal can be detected by the presence of lamellar or prismatic elongated fragments of nacre (Fig. 4.29). Each of these fragments is composed of long parallel columnar crystals of aragonite embedded in a protein matrix. Each of these nacre fragments, or crystals, is usually strongly birefringent when observed using polarized light. The same is true for oyster powder (Figs. 4.30 and 4.31).

Furthermore, large particles of shell from oyster (≥150 µm) also have other distinguishing features. They have long straight cracks crossing each other on the shell particle surface and elongated needle-like crystals spreading over different directions.

Squid meal mainly is characterized by the presence of numerous chitin remains originating from the mollusk's pen, beak, and tentacle suckers. Chitin fragments are fairly distinguishable. They appear as flat angular particles, lightly colored with often-visible successive deposits of layers (like roof tiles). Sometimes plate patterns like those from fish scales are present. These fragments (Fig. 4.32) also have numerous cracks at their surfaces, but unlike fish scales, they do not stain with Alizarin Red.

Acknowledgement

Micrographs with engraved logo are delivered by courtesy of the Community Reference Laboratory for Animal Proteins in feedingstuffs (CRL-AP) at Walloon Agricultural Research Centre, Gembloux, Belgium.

5 Detecting Animal Products in Feeds and Feed Ingredients

Introduction

For purposes of organization, this chapter is divided into six main categories:

- Background
- Value of Feed Microscopy
- Sample Preparation for the Detection of Processed Animal Products
- Examination
- Characteristics of Animal Products, and
- Interpretation of Observations and Limitations of Microscopy.

This chapter was written with North American and European collaboration. Although the methodologies used to detect animal products in feeds and feed ingredients are similar between both areas, there are some differences to note. These differences are primarily due to the various legislations of the individual jurisdictions between and within each area regarding processed animal products (PAPs) in feeds and feed ingredients.

The most significant difference between North America and Europe is that the method of analysis is written into the European Union (EU) legislation, whereas, this is not the case in the United States or Canada. The more prescriptive EU methodologies are highlighted in blue boxes within this chapter.

It is of utmost importance that microscopists using this manual are aware of and understand the legislation(s) of their country and how this could affect the manner in which feeds and feed ingredients analysis for the detection and identification of PAPs in feeds and feed ingredients is conducted.

Background

During the last revision of the *AAFM Manual of Microscopic Analysis of Feedstuffs* in the early 1990s, a significant event affecting animal feeds and feed ingredients, and the analysis of these products, was unfolding in the world. This event was the emergence of bovine spongiform encephalopathy (BSE) or "mad cow disease." This disease was first recognized in Great Britain in 1986 (WHO, 2002). Since that time, the number of BSE cases steadily rose in Great Britain and throughout the United Kingdom (UK) until it reached its peak in 1992. That year there was a total of 37,280 confirmed cases of BSE in the UK, 36,682 of which were in Great Britain (OIE, 2007). Over the years, this disease began appearing in numerous other European countries, Japan, and North America. The first case in North America was detected in Canada in 1993 in an animal imported from the UK (CFIA, 2009). The first indigenous case of BSE in North America was detected in Canada a decade later, in 2003 (CFIA, 2009). A short period of time later, the United States of America (USA) reported their first case of indigenous BSE in 2005 (OIE, 2010).

BSE is categorized in a group of diseases collectively known as transmissible spongiform encephalopathies (TSEs) or prion (*prēŏn*) diseases (Doherr, 2007; Prusiner,

1998). TSEs are progressive neurodegenerative diseases that affect the nervous system of susceptible mammals causing loss of physical and mental functions and inevitably death (Prusiner, 1998). Several mammalian species have proven to be susceptible to TSEs, including: cows (BSE), sheep (scrapie), deer and elk (chronic wasting disease; CWD), mink (transmissible mink encephalopathy; TME), etc. (Doherr, 2007; Prusiner, 1998).

Humans are not exempt from TSEs and several have been identified, including: Kuru, Gerstmann-Sträussler-Scheinker syndrome (GSS), Fatal Familial Insomnia (FFI), Creutzfeldt-Jakob disease (CJD), and variant Creutzfeldt-Jakob disease (vCJD) (Doherr, 2007; Prusiner, 1998). These diseases are 100% fatal and currently there are no cures or treatments (Prusiner, 1998).

Although there still is some debate in the scientific community (Telling et al., 1995), it is widely believed that the infective agent of TSEs is a misfolded prion protein (commonly abbreviated PrP^{SC} for *Pr*ion *P*rotein$^{\mathrm{DiSease\ Causing}}$) (Doherr, 2007). This is unlike other more "classical" infectious diseases that are caused by viruses, bacteria, protozoa, parasites, etc. Susceptible mammals, including humans, naturally produce prion proteins in the native form (PrP) (Doherr, 2007). The function of this protein is unknown (Prusiner, 1998), but it does appear to be produced and degraded in a natural equilibrium as are other proteins. It is believed that when a native PrP protein comes into contact with the infectious PrP^{SC} protein, the PrP protein becomes misfolded and is converted into PrP^{SC} (Doherr, 2007; Prusiner, 1998). The infectious PrP^{SC} protein is not susceptible to natural degradation and begins to aggregate into plaques which accumulate in specific tissues of the body, most importantly in the central nervous system. As the disease progresses, small holes, or vacuoles, begin forming in the brain causing it to appear sponge-like (Doherr, 2007), giving rise to the term "spongiform." Unlike most other infectious agents, the incubation period of this disease can be months or years after initial exposure before symptoms begin to appear, as opposed to hours or days. In cattle, BSE has a mean incubation period of 4–5 years (WHO, 2002). This long incubation period creates significant challenges to scientists studying this disease.

The most widely accepted route of transmission of BSE is through feedstuffs (Doherr, 2007; Prusiner, 1998). The exposure route to this disease is created by feeding cattle processed animal proteins that contain infective material. One theory is that BSE originated by feeding cattle feed that contained sheep protein products tainted with scrapie, a disease that has been recognized in sheep for centuries. Once this disease crossed the species barrier and became more readily infective to cattle, BSE was born. Another theory suggests that BSE was a spontaneous occurrence. Yet another theory suggests that BSE was derived from CJD (Colchester & Colchester, 2005). No matter the origin of this disease, it is believed that BSE was allowed to proliferate in the cattle population as the infective agent was recycled back into the cattle feed chain through the use of bovine protein products (e.g., bovine meat and bone meal). The cattle consumed the infective material, became infected, the infective agent amplified, the cattle were slaughtered, and then the unknowingly infected carcasses were used to produce more infective bovine protein products. In the past, this type of cannibalistic recycling of processes animal proteins was a common and acceptable practice in the ruminant feed industry to both utilize remaining animal carcasses, thus reducing waste, and to increase the nutritional content of feeds in a cost effective manner. In addition to this type of recycling, the rendering process used to produce meat and bone meals (MBMs) in the UK changed after 1970 and into the 1980s when a solvent extraction step used during rendering was phased out (Wilesmith, 1991). During this time, the use of the solvent extraction process and a second steam heating treatment of MBMs were eliminated. These changes potentially allowed sufficient quantities of the infective agent to remain active and may have contributed to scrapie crossing the species barrier and/or allowed BSE to be transmitted. Unfortunately, the recycling and change in rendering practices may have created ideal conditions for this disease to proliferate.

BSE not only has an impact on cattle health, but human health as well. In 1996 a new form of CJD was reported in Great Britain; variant Creutzfeldt-Jakob disease (vCJD) (WHO, 2002b). This disease is thought to be transmitted to humans through the consumption of BSE-tainted beef products (WHO, 2002b).

As the epidemiological links between BSE and feed, and BSE and vCJD increased, governments around the globe began placing safeguards to protect against these diseases. In 1988, the feeding of ruminant protein products to animals was banned in the UK. Six years later, in 1994, the European Union (EU) banned the feeding of all mammalian products to ruminants by Commission Directive EC/381/91 (European Commission, 1994). This feeding ban was revised in 2001 and a total feed ban prohibiting MBMs in all farm animal feeds in the EU was imposed to help protect against cross contamination of feeds and accidental feeding of cattle with feeds formulated for other species that contain prohibited products (European Commission, 2001). In 2005, the European Commission adopted a checklist for TSE. The document, referred to as "The TSE Roadmap," presents different scenarios for possible future changes in EU legislations concerning global BSE concern while continuing to give high priority to food safety and consumer protection (European Commission, 2005). One of the objectives is to ease, under well defined conditions, some aspects of the total ban of animal proteins actually in force. In this respect, the document states the urgent need for method development. In 2008, the EU made a first step in easing the total feed ban by allowing the reintroduction of fish meals in milk replacers for young ruminants (European Commission, 2008).

Numerous other countries, such as the United States and Canada, initially responded to the threat of BSE by placing import restrictions to control the movement of potentially infective materials from countries with BSE. In 1997, both the United States and Canada banned the feeding of specific animal products from all mammalian species (excluding horse and pig) to ruminants in their own distinct, but similar feed bans. Canada further enhanced this feed ban in 2007 to exclude the use of specific risk materials (SRMs)—the skull, brain, trigeminal ganglia, eyes, tonsils, spinal cord, and dorsal root ganglia—in animal feeds, pet foods, and fertilizers (CFIA, 2007).

Value of Feed Microscopy

Feed microscopy is an important tool to verify compliance with the various feed bans in place around the world. In some instances feed microscopy can be the sole method used to determine compliance, while in other countries it must be used in conjunction with other methods. For example, countries that have complete feed bans (i.e., all animal products are prohibited in feeds) can almost solely rely on feed microscopy as it is relatively easy to detect even minute amounts animal product contamination. Nonetheless, other methods [e.g., DNA-based (PCR), immunoassay, near infrared microscopy (NIRM)] can be complementary to microscopy, especially when prohibited ingredients fail to be detected by feed microscopy alone, when additional information on the origin of the contamination is requested or when processing treatment of animal products, as those used in EU, lead to the destruction of many microscopic markers. In countries that do not have a complete feed ban (i.e., only specified animal products are prohibited in specified feeds), animal product contamination must not only be detected, but the feed ingredient and its species of origin must also be identified. Feed microscopy can identify the product, but the identification of the species of origin of the product is more challenging. This is especially true when examining feeds with relatively small amounts of contamination, e.g., <0.5% (w/w), or feed processed by high temperature-pressure treatment.

In both the United States and Canada it is prohibited to feed ruminants with most mammalian products, but all horse and pig derived products are exempt from these bans. Using microscopy, it is relatively simple to determine if mammalian products are present in a feed by examining bone fragments (should enough be present). However, unless a feed is contaminated with a relatively large amount of prohibited product, e.g., ≥0.5% (w/w) bovine meat and bone meal; which is equivalent to 5 kg of prohibited material per metric ton of feed, it may be difficult to detect hair fragments and determine if the mammalian product is derived from a prohibited species or not. In such cases, other methods are required to determine the species of the source of contamination, such as DNA-based methods. This being said, it is also important to note that DNA-based (and other) methods must also rely on feed microscopy. DNA-based methods have been shown to detect minute quantities of animal products in feeds (i.e., 0.001% w/w) and determine the species of origin. However, current DNA-based methods cannot differentiate the tissue(s), and thus type(s), of animal product present to determine if it is prohibited or not. For example, DNA-based methods would indicate the presence of a prohibited species if bovine blood meal or milk derivates were present, although those ingredients from all mammals are permitted in ruminant feeds. These methods must rely on feed microscopy and vice versa.

This chapter will outline how to detect, and possibly identify, relatively small amounts of ingredients of animal products. For more detailed descriptions of specific animal products, see Chapter 3 Feed Ingredients of Animal Origin, and Chapter 4 Feed Ingredients of Marine Origin.

Principle

Feed microscopy can be used to detect animal products in feeds, even when present in minute amounts, e.g., 0.001% (w/w). However, the smaller the amount of processed animal product (PAP) present in a feed, the more difficult it becomes to (A) detect and (B) identify.

Both detection and identification of PAPs is based on detecting and examining morphological features of markers specific to animal products. These markers include bone and cartilage fragments, muscle fibers, hairs and bristles, feathers, blood, milk, fish bones, scales, etc.

If any of these markers are detected, then animal products are present in the feed. In most cases, several markers are required to identify the specific PAP(s) present. For example, to identify porcine meat and bone meal in feed, muscle fibers, terrestrial bone fragments, and porcine hair must all be detected. Thus, detecting PAPs is generally simpler than identifying them.

When a feed is examined to determine if PAPs are present, it is important to know where to look in the feed to find the markers. Proper sample preparation is essential. Although a general sample preparation is described in Chapter 1 Methods of Agricultural Microscopy, this chapter is specific to the preparation of feedstuffs for the detection of trace amounts of PAPs.

The method described in this chapter constitutes a guideline for Northern American countries. EU member states must follow a similar method for official controls described in Annex VI of Commission Regulation 152/2009. Blue boxes in the chapter allow referring to the EU method and its implementation. These boxes also focus on differences among respective methods.

Sample Preparation for the Detection of Processed Animal Products

The following sample preparation has to be performed on feeds or ingredients with low moisture content. Check for a moisture content below 14%. In case of higher values of moisture, dry the sample in an oven before starting the preparation protocol.

A sample should be reduced in size to approximately 20 g while ensuring a representative subsample is obtained. This can be accomplished by using a riffler, quartering the sample (see "Preparing the Sample for Inspection" in Chapter 1 Methods of Agricultural Microscopy), or by other acceptable techniques. Samples that consist of feed pellets (In such cases where a feed consists of feed pellets mixed with other ingredients, e.g., a complete feed, enough feed pellets (i.e., ≥5 g) should be removed, crushed, and treated as an independent subsample.) should be ground using a mill or crushed using a mortar and pestle with enough pressure to break apart the pellet with minimal damage to the individual ingredient particles. The subsample should then be mixed and quartered to obtain two ~10 g fractions. The first fraction is to be examined without any further preparation and will be referred to as the "as is" fraction. The second fraction is to be processed further by floatation.

Floatation

A basic floatation is performed on the subsample to separate the organic constituents of a feed from the denser mineral/bone constituents. This simple procedure concentrates specific animal products into two distinct fractions, making it easier for the microscopist to detect them (should they be present).

To perform a basic floatation, set up a simple filtration system in a chemical fume hood. Choose a funnel large enough to contain the volume of the second ~10 g subsample. Place a filter paper (e.g., Whatman #1) into the funnel and place a container under the funnel to capture the chloroform ($CHCl_3$) [Other solvents can be used—e.g., carbon tetrachloride (CCl_4), tetrachloroethylene (C_2Cl_4 or TCE)] waste.

Place the ~10 g subsample in a ceramic dish large enough to contain approximately twice its volume. Next, enough $CHCl_3$ is added to cover the subsample while allowing the less dense material to float. The subsample is stirred for approximately 60 s and then allowed to settle for 1–2 min. Skim off the organic feed constituents that are floating and place it into the filter paper. Take care not to disturb the denser materials at the bottom of the dish. Once the denser materials are visible, gently pour the $CHCl_3$ and organic materials into the funnel. Add additional $CHCl_3$ if necessary, but take care not to disturb the dense material. Once most of the organic materials have been separated from the denser materials, allow both fractions to air dry. The organic material that ended up in the funnel is referred to as the "flotate." The dense material that remained in the ceramic dish is the "sediment." The majority of fats and oils will have been washed away in the $CHCl_3$.

NOTE: Be extremely careful for cross-contamination during air drying. Because of solvent release, drying of the fractions is often carried out under the fume hood with an almost closed front panel. Never close this front panel but leave it open wide. Otherwise, the air flow generated by the extraction engine of the fume hood might transfer particles from one vial to the others.

Sediment—Staining with Alizarin Red

The sediment can be dyed with the histological stain Alizarin Red to help differentiate bone fragments from minerals present in the sample. This product will stain bone fragments a bright pinkish-red color, making them more easily distinguishable from similar looking minerals (e.g., limestone). However, it must be noted that Alizarin Red is not a bone-specific stain. The dye is known to be a chelator of calcium that stains more specifically hydroxyapatite—the major mineral component of bone and dentine. The dye is believed to react principally with the molecules of tricalcium phosphate composing the hydroxyapatite. Therefore, it does stain some minerals containing calcium (e.g., mono-, di-, and tricalcium phosphate) or other calcium containing structures (e.g., fish scales) in a similar color. This stain is used to help highlight the presence of bone fragments, but other histological characteristics must be observed to confirm if a fragment is indeed bone (see Characteristics of Animal Products—Bone below). Alizarin Red solution is prepared in the following manner.

0.2% (w/v) Alizarin Red Solution

Add 2.5 mL 1N HCl to distilled H_2O to make a 100 mL solution. Dissolve 200 mg Alizarin Red in this solution. Store in a glass bottle.

Staining procedure:

1. Transfer the sediment into a glass test tube (e.g., 18 mm × 150 mm).
2. Wash twice with ~5 mL 95% ethanol. [All washes are as follows: vigorously mix (e.g., Vortex) for >10 s, allow settling for >1 min, decant liquid waste, and retain sediments.]
3. Add enough sodium hypochlorite (Reagent grade sodium hypochlorite with 10–13% available chlorine) to cover sample (usually 1–2 mL).
4. Allow reaction to occur for 10 min.
5. Add distilled H_2O to fill test tube 2/3 full.
6. Allow materials to settle for 2–3 min.
7. Decant liquid with suspended particles and retain sediments.
8. Wash twice with ~10 mL distilled H_2O.
9. Add sufficient amount of Alizarin Red solution to cover the sample (e.g., 1–2 mL).
10. Gently mix by swirling for at least 30 s.
11. Immediately wash twice with ~5 mL 95% ethanol.
12. Wash once with ~5 mL acetone.
13. Allow to air dry or store in an oven (60–90C) to accelerate drying.

NOTE: If the sample is to be examined for all ingredients (i.e., not just animal products), perform this staining procedure after the sediment has been examined and the mineral constituents have been identified as many minerals (e.g., salts) will be removed from the sample by the washing steps.

In the EU, a sample of at least 50 g of material is required. Feed pellets may be pre-sieved to obtain two fractions. The two fractions obtained have to be considered and analyzed as two independent samples. Mortar and pestle crushing (or the use of a grinding mill) is also recommended to achieve the appropriate fine texture. After grinding, two subsample fractions of at least 5 g are taken for further examination. One for the sieve fractions analyses (comparable to the "as is" fraction); one for the concentrated sediment analysis.

A floatation to segregate sediment from flotate is performed but using tetrachloroethylene (TCE) (with a density of 1.62) instead of chloroform. Perform floatation as follows.

In a fume hood, place a 250-mL conical separation funnel—as chosen by most EU member states—on a rack over an open beaker. Below the stopcock of the separation funnel, install a funnel placed on a container to capture the TCE waste. Place a filter paper (e.g., Whatman #1) into the funnel to collect the sediment.

Transfer at least 5 g of feed into the separation funnel. Add 50 mL of TCE into the separation funnel. Close the vial and shake vigorously for 60 s. Then add a second 50-mL aliquot of TCE alongside the wall of the separation funnel to clean it from all adhering feed fragments and to bring them to the bottom. Allow settling for 5 min without stirring; the two phases will separate nicely. Recover the sediment on the filter paper into the funnel by opening the stopcock. Close immediately after the sediment has passed. The sediment must be air dried before use. The flotate can also be examined—although not specified in the official method text. For isolation of the flotate, empty the separation funnel from its TCE with the stopcock and close the stopcock when it is almost emptied. Turn the separation funnel bottom-up over a large glass petri dish. Open the stopcock, most of the flotate will fall into the petri dish. Rinse the wall of the separation funnel with a stream of TCE through the tube opening. Allow the flotate to air dry under the fume hood. The segregation of the bone/mineral constituents from the other organic fraction is almost absolute as the density of TCE is higher than that of chloroform (d = 1.48). TCE has a comparable fat washing action as chloroform. All in all, advantages of the EU method are (1) a better separation of organic vs. mineral/bone fractions and (2) a production of sediment with a higher bone ratio.

Alizarin Red staining of the sediment is also recommended in the EU official method.

Examination

After the sample preparation is complete, the ~20 g subsample has become three separate fractions; the "as is" fraction, the flotate, and the sediment. These fractions are handled and examined in different manners as particular animal product markers will be more concentrated, and thus more easily detected, in one fraction as opposed to another.

The flotate is examined with both the stereoscope and the compound microscope. The same is true for the sediment. However, different animal product markers will be present in each of these fractions. If sufficient animal product markers have been detected in the flotate and sediment, a complete examination of this fraction may not be necessary. The "as is" fraction is generally the last fraction to be examined and is usually only examined for the presence of milk products.

The EU method also recommends the examination of the different fractions.

The "as is" fraction must be sieved with a square mesh sieve of maximum 500 µm. This obligatory sieving step produces two sieved fractions.

1. a coarse-sieved fraction to examine with a stereoscope
2. a fine-sieved fraction to examine with a compound microscope

The sediment can also be sieved on a 500-µm mesh sieve and delivers two subfractions.

1. a coarse sediment
2. a fine sediment

Both fractions of the sediment need to be analyzed with a stereoscope and a compound microscope.

The EU protocol does not refer officially to the use of the flotate. Although many European laboratories routinely examine this fraction as it is partly defatted by the TCE floatation and therefore more suitable for fragment separation—especially for observations under the stereoscope.

Flotate—Examination with a Stereoscope

A number of animal product markers can be detected when examining the flotate with a stereoscope, including: hair fragments, feather fragments, gelatin, and muscle tissue.

In most cases, the flotate will be approximately 8–10 g. Depending on its general color and appearance, it is spread onto a white or black square piece of glass in order to achieve the best possible contrast (sheets of white or light blue paper may also be used). Spread the flotate in a sufficiently thin layer that allows examination of all particles. Alternately, the flotate can be quartered again until it is sufficiently small enough to cover the bottom of a plastic petri dish (e.g., 2–5 g). If sufficient animal products markers are detected in this smaller fraction, the remainder of the flotate need not be examined.

The flotate should be examined at a 10× magnification. To ensure that it is examined in its entirety, the sample should be scanned starting at one edge of the sample. The sample should then be moved vertically in the field of view until the bottom of the sample is reached. The sample is then moved horizontally by one field of view and again vertically until the top of the sample is reached. This scanning pattern is repeated until the entire sample is examined. To further ensure the entire flotate is examined, particles in the field of view can be pushed from the field of view with a microspatula while this scanning pattern is followed. For example, if the examination of this fraction commences at the top left-hand edge of the sample, the particles in the field of view should be moved to the left as the examination proceeds.

During the examination, suspect animal products should be pulled from the sample with forceps for further examination with the compound microscope. Examination of suspect muscle, feather, and hair fragments with the compound microscope is required to confirm their identity as often there are non-animal products (e.g., various plant products, synthetic fibers, etc.) that can be mistaken as animal markers. Distinguishing between animal product markers and similar looking products using the stereoscope can be difficult, if not impossible in some cases. However, this distinction is generally quite clear when the suspect specimen is observed with the compound microscope.

Flotate—Examination with a Compound Microscope

Only a small portion of the flotate is examined with the compound microscope. Although the sample size is small, detection of even minute amounts of animal products in a feed can be accomplished if the sample is properly prepared and mounted on a slide. Muscle fragments, feather fragments, and occasionally hair fragments can be detected with the compound microscope.

Other than suspect fragments removed from the flotate during the examination with the stereoscope, only the fine (i.e., ≤0.5 mm) fragments are examined using the compound microscope. Fine fragments can be isolated from the flotate by using a sieve with the appropriate 500-µm mesh size. Once the fine fragments have been isolated, place a microspatula tip full of this material into a small glass test tube (e.g., 10 mm × 75 mm). Add ~15 drops of a 60% (w/v) aqueous chloral hydrate solution and gently bring to a boil for a few seconds using an alcohol lamp. The chloral hydrate solution will hydrolyze starch and act as a clearing agent which will provide a greater clarity of the sample. Once the solution has cooled, place ~12 drops onto a slide and gently cover it with a large coverslip (i.e., 22 mm × 60 mm) while taking care to avoid trapping air bubbles. Sometimes the details of larger particles (i.e., ≥0.5 mm) from the flotate have to be checked and hence need to be observed under the compound microscope. In these cases, the use of slides with concavities (e.g., "hanging drop" slides) is recommended over classical flat glass slides.

Scanning of the prepared slide for animal products should be performed at a magnification of at least 100× using transmitted brightfield light. Suspicious particles can easily be observed at a greater magnification to verify their identity. Start the scanning process at one of the corners of the coverslip. First move the slide vertically in the field of view while scanning until the other corner of the slide is reached. Move the slide horizontally by one field of view, then move the slide vertically while scanning. Repeat this pattern until the entire slide has been scanned. The slide should be examined right after it is prepared, but can be examined up to a few hours later as long as the chloral hydrate solution has not evaporated.

Sediment—Examination with a Stereoscope

Unless examining a mineral supplement or other premix, the sediment will usually be ≤2 g. It is recommended that the sediment be examined using a dark background. A dark piece of paper will suffice, but a black or cobalt blue piece of glass (~10 cm × 10 cm) is recommended. Dump the entire heavy fraction from the ceramic evaporation dish (or from the test tube if the Alizarin Red staining procedure was performed) onto the dark background such that there is a single layer of fragments (The very fine sediment fragments will form a thin layer at the bottom of the ceramic dish or test tube. Do not disturb these particles, as they will be used to prepare a slide to be examined with the compound microscope.). As with the flotate, it is recommended that the sediment be examined at a 10× magnification. Increase the magnification as necessary if the sediment is very fine.

During the examination, suspect animal products should be pulled from the sample for further examination. Once the microscopist is familiar with the appearance of bone fragments, they are usually readily identifiable. However, if there is a relatively small amount of bone fragments present and the microscopist is unsure of the identity, the suspect bone fragment should be crushed on a microscope slide, mounted, and observed with the compound microscope (see Sediment—Examination with a Compound Microscope). Additionally, spot tests can be used to help confirm the identity (see Table 9.D—Spot Tests for Animal Products in Chapter 9 Microchemical Spot Tests).

Sediment—Examination with a Compound Microscope

As with the flotate, the fine (i.e., ≤0.5 mm) fragments of the sediment are examined using the compound microscope. Fine fragments can be scraped from the bottom of the ceramic evaporation dish used in the floatation process, knocked from the test tube wall from the Alizarin Red staining procedure, or isolated from the sediment using a sieve with the appropriate mesh size. The last option is preferred, as it will guarantee better homogeneity of the sediment subsample used for slide preparation. As with the flotate, larger particles (i.e., ≥0.5 mm) from the sediment sometimes need to be observed under the compound microscope. Again, the use of slides with concavities (e.g., "hanging drop" slides) is recommended over classical flat glass slides.

Permanent or non-permanent slides can be prepared depending on the record keeping requirements of the laboratory. If a permanent slide is prepared, it can be examined at any time, and is useful should one wish to obtain a second opinion from a colleague. If, however, a non-permanent mounting medium is used, the slide must be examined immediately as some of the structural features of bone (e.g., lacunae) may become filled with the mounting medium in a short period of time (e.g., <5 min) causing them to lose contrast and become difficult to distinguish from the rest of the bone fragment. For this reason, mounting media to use for non-permanent slides should have a high viscosity.

Two non-permanent mounting media are recommended: paraffin oil and pure glycerol. Paraffin oil is a mixture of fluid hydrocarbons that have varying chemical properties and quality that vary among producers. For this reason, preference should be given to glycerol, which is a pure chemical compound.

For permanent slides, the recommended mounting medium is Norland Optical Adhesive 65. Advantages of using this resin have been described by Veys and Baeten (2010). It is a clear and highly viscous resin with a refractive index similar to glass that can be rapidly cured using UV light ($\lambda = 365$ nm). The refractive index similar to glass allows the light to pass through the slide and specimen in a relatively straight path, which produces a very clear image of the specimen. The high viscosity and the rapid curing time of this mounting medium prevents the structural features of bone (e.g., lacunae) from filling with the mounting medium, thus permanently retaining the contrast of these characteristics.

Slides are prepared as follows. A few drops of mounting medium are placed on a dry clean microscope slide. The fine sediment fragments are then placed on the mounting medium using a microspatula and stirred gently to obtain a uniform spread of the particles. Avoid the formation of air bubbles while stirring. Finish the slide preparation by laying down a coverslip (e.g., 22 mm × 22 mm) over the material while taking care to not trap bubbles. The entire slide preparation is examined as described in "Flotate—Examination with a Compound Microscope."

Within the EU, the slide preparation on the different fractions obtained (see previous box) is comparable to Northern American guidelines. Nevertheless, the microscopic detection of PAPs has to be performed on a minimum number of slides.

- at least one slide of the fine-sieved fraction to screen systematically
- at least one slide of the fine sediment

If from this first analysis, fish is detected, then extra analysis is required.

- at least two additional slides of the fine sieved fraction will be examined
- the total sediment fraction must be analyzed

Characteristics of Animal Products

Blood

Blood concentrates in the flotate are generally more easily detected with the stereoscope than with the compound microscope. Suspect blood particles should be removed from the flotate, placed onto a dry microscope slide, gently crushed, and covered with a coverslip (The fine [i.e., <0.5 mm] fragments from the flotate can be screened for blood in the same fashion.). The particles should then be placed in the field of view of the compound microscope using transmitted brightfield light. Once in focus, place a few drops of the 3,3′,5,5′-tetramethylbenzidine (TMB) and hydrogen peroxide solution (as described in Chapter 9 Microchemical Spot Tests—Method XX) next to the coverslip so that it is drawn under the coverslip by capillarity. Observe the reaction immediately. A positive reaction for blood will produce a blue/green color that will move in the direction of the reagent front and the formation of oxygen bubbles around the suspect blood particles (Fig. 5.1).

This reaction is relatively specific to blood, as it does not occur with pure muscle fibers or plasma powder (There is formation of a blue/green color when blood plasma is tested, but there is no oxygen bubble formation.). The blue/green color is the TMB reacting with the iron present in blood, and the oxygen bubbles are formed by the decomposition of H_2O_2 into oxygen and water by the catalase enzyme present in blood. Nevertheless, a positive control using a fragment of blood meal from a known reference collection should be used to verify that the reagent is working properly and to allow the microscopist to observe a proper positive reaction for comparison.

Stereoscopic Features. Blood is primarily present in a feed as blood meal. However, small amounts of blood can also be present in meat meals, meat and bone meals, poultry meals, etc. No matter the source, blood is fairly distinctive by its dark burgundy to black

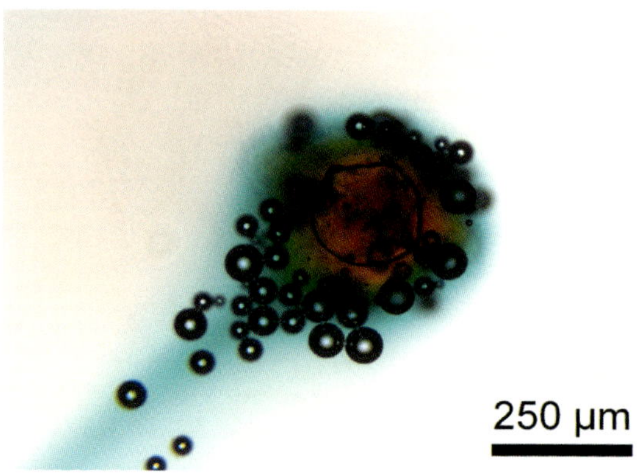

Fig. 5.1. Positive spot test for blood observed using a compound microscope. In this case, the front of the reagent was moving from the top-right to the bottom-left of the field of view. The blue/green color produced by a positive reaction follows this reagent front. The circles with the dark outline are the oxygen bubbles trapped underneath the coverslip. © Her Majesty the Queen in Right of Canada (2010).

color. Depending on the processing of the blood meal (e.g., drum dried heat treated or spray dried cold treated), it can be present in a feed as coarse, irregular, rough fragments (up to 2–3 mm) when drum dried (Fig. 5.2), or as small, powdered, dark spheres (around 50 µm in diameter) or dark and glossy fragments (up to 1mm) when spray dried (Fig. 5.3). When present as spheres, multiple layers similar to that of an onion can often be seen when the fragment is bisected.

Histological Features. Blood present in feed is generally not detected with the compound microscope, as the minute particles are amorphous in shape. If, however, amorphous globular particles with a translucent light wine red to ruby red color and an air-filled center, similar to a vacuole, are observed when examining the fine particles of the flotate, then this fraction should be carefully observed for blood with the stereoscope and/or the fine fragments should be screened using the histochemical spot test for blood as describe previously.

Fig. 5.2. Drum dried blood meal observed using a stereoscope. © Her Majesty the Queen in Right of Canada (2010).

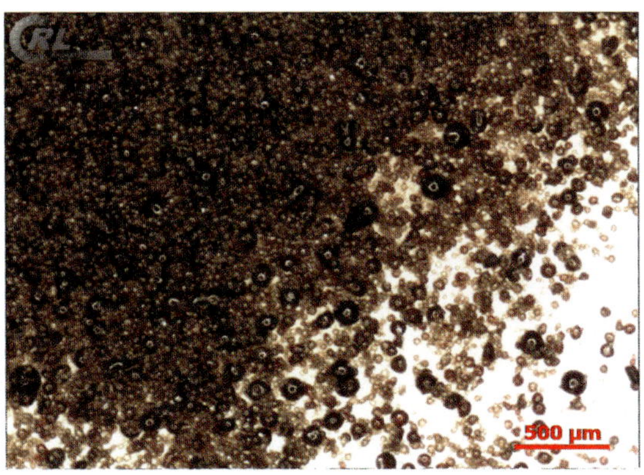

Fig. 5.3. Spray dried blood meal observed using a stereoscope. Micrograph by courtesy of CRL-AP.

Potential Misidentification Traps. Care should be taken not to confuse feedstuffs that are dark in color with blood. For example, the cottonseed hull fragments of cottonseed meal or the dark fragments of canola meal could lead a microscopist to erroneously conclude that blood meal is present. Confirmation of blood using the spot test described in Chapter 9 Microchemical Spot Tests (Method XX) is highly recommended.

Milk Products

Milk products (e.g., milk powder and lactoserum powder) are arguably one of the most difficult animal products to detect in feeds via microscopy, especially when present in pelleted/extruded feed (the high pressure steam extrusion process makes them virtually undetectable by microscopy) and/or when present in minute quantities.

When examining a feed for milk products, the "as is" fraction should be examined as the chloroform used in the floatation process will remove much of the milk products from the sample.

Stereoscopic Features. Milk products generally appear as a fine white, yellow, or tan powder depending on its fat content. Skim milk powder will be almost pure white, whereas, fat enriched lactoserum (whey powder) will appear yellowish.

To help identify milk powders, place two small portions of the fine particles from the "as is" fraction on a microscope slide. Place a drop or two of water on the first portion and observe against a dark background. Milk powders may appear as glossy white to pale blue particles and can have a cloudy halo or stream emanating from it (Fig. 5.4). Next, place a

Fig. 5.4. Milk powder particles in water observed using a stereoscope. Whitish streams, like comet tails, are visible. Micrograph by courtesy of CRL-AP.

drop or two of an iodine solution (Dissolve 1 g KI in 50 mL water and add 0.5 g I_2 while stirring. Store in an amber dropper bottle.) on the second portion of fines and observe against a black background. Milk powders may appear as glossy white to light yellow particles while starch particles present will react with the iodine and turn dark purple.

Histological Features. Dried milk products may be detected as relatively small (e.g., <50 µm) particles with a light yellow tinge that have a somewhat amorphous/circular shape with vacuole-like structures in the center (Fig. 5.5).

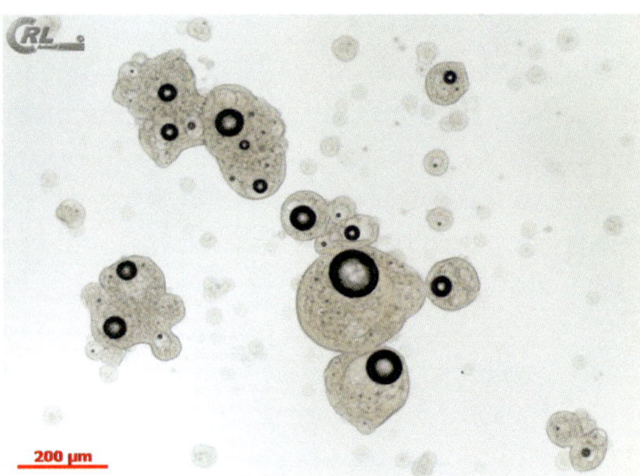

Fig. 5.5. Milk powder particles observed using a compound microscope. Micrograph by courtesy of CRL-AP.

Large lactose crystals may be present in milk powder but are more likely found in lactoserum powder. Those crystals appear as axe head (α-lactose) (Fig. 5.6) or kite-like shaped (β-lactose). When illuminated with polarized light, these crystals display bright purple, green, and blue birefringence in a concentric pattern that follows the contour of the crystal (Fig. 5.7). Other more discrete types of lactose crystals (depending on the respective concentrations of α and β isomers of the sugar or its amorphous form) are present in milk powder. They are regularly present and appear as a bright, thin lining surrounding the milk

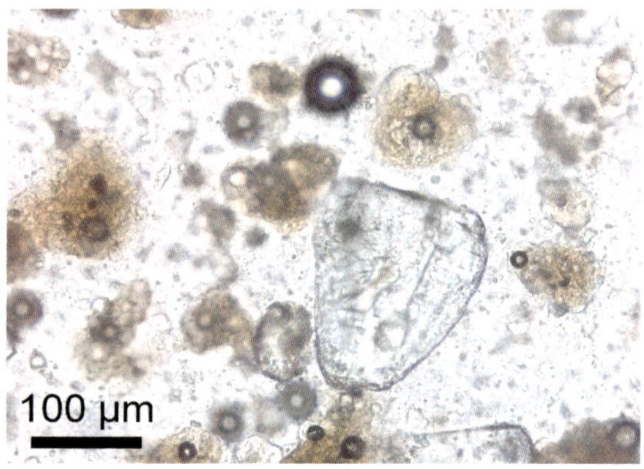

Fig. 5.6. α-lactose particle of a lactoserum (whey) powder observed using a compound microscope with brightfield light. © Her Majesty the Queen in Right of Canada (2010).

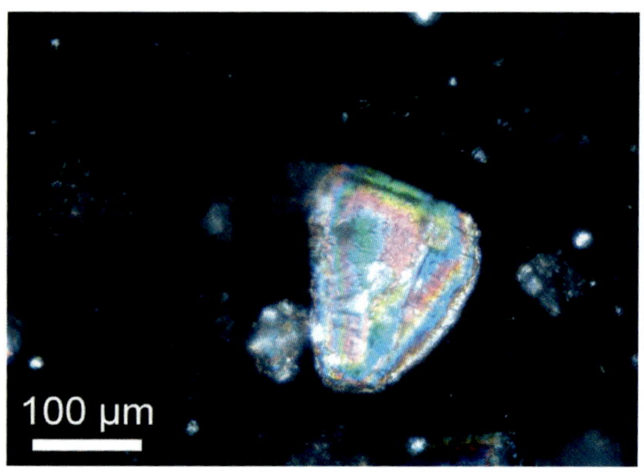

Fig. 5.7. The identical α-lactose particle as the previous figure. observed using a compound microscope and polarized light rather than brightfield light. © Her Majesty the Queen in Right of Canada (2010).

globules when observed with polarized light. Note that those lactose crystals are freely soluble in aqueous mounting media; thus, if slide mounting medium consists of water, they will no longer be visible. Mounting with Norland Optical Adhesive 65 or even pure glycerol is recommended.

Large fat globules may also be present especially in whole milk powders. These fat globules appear as perfectly circular oil droplets that generally are located at the surface of the mounting medium. They do not display birefringence when observed using polarized light, and classical Sudan dyes may be used to make them more easily distinguishable.

Potential Misidentification Traps. Milk products are similar in appearance to other light colored powdered products. Starch particles from sources such as ground grains may be misidentified as milk products. The iodine test described above will allow for the differentiation with the stereoscope. When observed with the compound microscope, starch particles do not have the vacuole-like appearance of milk. Under polarized light, starch particles have an intense birefringence and may display a maltese cross, if the starch has not been gelatinized.

Muscle Tissue

Muscle tissue can be very difficult to detect and identify using a stereoscope, especially when compared to the relative ease of muscle tissue detection and identification with the compound microscope. Muscle tissue detection is not impossible if only a stereoscope is available; however, confirmation with a compound microscope is highly recommended.

Most muscle tissue from meat and bone meals is finely ground and will concentrate in the fine fragments of the flotate that is examined with the compound microscope.

Stereoscopic Features. Larger muscle tissue fragments are generally cylindrical in shape and caramel in color. These large fragments are usually not much longer than a few millimeters in length and less than half a millimeter in diameter. Most muscle tissue fragments are too fine to detect with the stereoscope.

NOTE: Harsher treatment of meat and bone meals, as used in the EU (i.e., 133C at 3 bars for 20 min), results in the production of meat and bone meals with an almost total absence of any muscle tissue fragments identifiable with a stereoscope. At higher temperature treatment, the histological features start to be affected, leading to some hardly recognizable structure observations.

Histological Features. Muscle fibers are relatively easily distinguishable, even at a magnification of 100×, and will be present either as individual muscle fiber fragments, or, less frequently, they can be present in fiber bundle fragments (Fig. 5.8). Fragments are usually transparent but can be found to have some faint yellow to orange coloration. They are almost always rectangular (i.e., cylindrical) in shape although long, thin strands are rarely present. Sometimes they present typical small but deep folds at their surface making them sausage like in appearance.

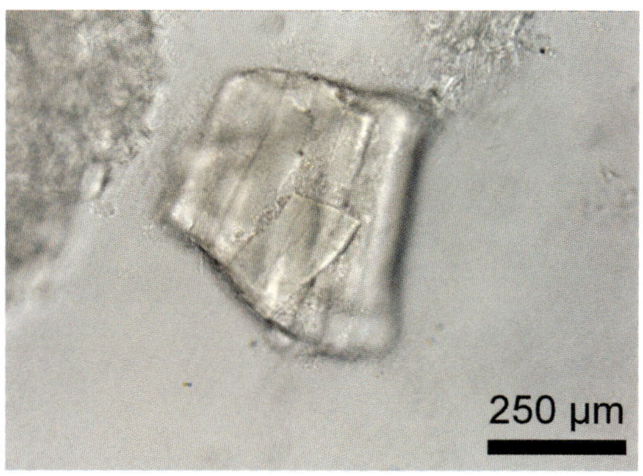

Fig. 5.8. Muscle fiber bundle observed using a compound microscope. © Her Majesty the Queen in Right of Canada (2010).

As muscle fibers primarily originate from voluntary muscles, the most characteristic feature is the presence of the striations of the muscle cells' sarcomers. The striations are seen in the muscle fragment as very thin dark parallel lines that are spaced a few microns apart (Figs. 5.9, 5.10). It may be necessary to observe a suspect muscle fragment at 200× or even 400× magnification to easily observe the striations. Special attention should be paid to the amount of contrast created by adjusting the aperture of the microscope condenser diaphragm. If there is not sufficient contrast, the striations, and indeed the muscle fragment itself, may not be visible.

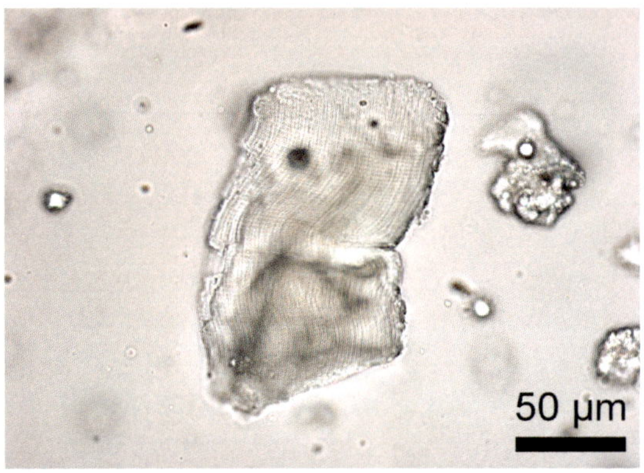

Fig. 5.9. A single muscle fiber fragment. The striations are visible as parallel, dark, thin lines running perpendicular to the long sides of the fragment. © Her Majesty the Queen in Right of Canada (2010).

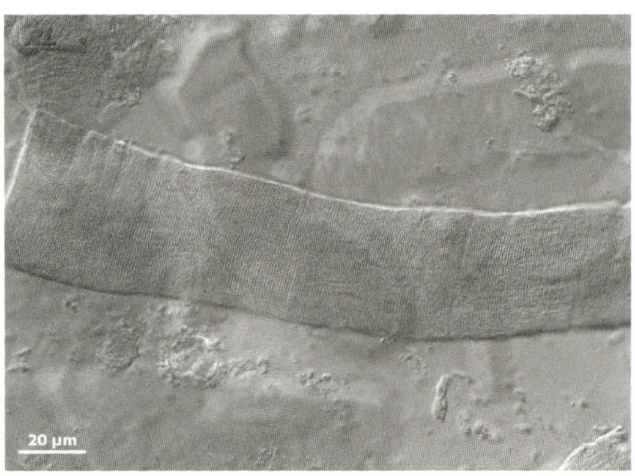

Fig. 5.10. A single muscle fiber fragment observed by DIC showing alternating artificial crests and grooves. Micrograph by courtesy of CRL-AP.

If the contrast is too high, muscle fragments may be obscured by other particles in the sample. Using differential interferential contrast (DIC) illumination is recommended as it enhances contrast of the striations, making them more easily visible while also enhancing the alternating crests and grooves (Fig. 5.10).

The striations of muscle tissue can be highlighted if the sample is mounted in an iodine solution (Dissolve 1 g KI in 50 mL water and add 0.5 g I_2 while stirring. Store in an amber dropper bottle.) as opposed to a chloral hydrate solution. The disadvantage to using an iodine solution is that starch present in a sample is not hydrolyzed and may interfere with the observations. Detection of muscle fibers may also be confirmed, when the striating aspect is missing or barely visible, by the Fehling's reagent (as described in Chapter 9 Microchemical Spot Tests—Method XXI). Fehling's reagent specifically stains the glycogen contained in the muscle cells pinkish-violet.

Potential Misidentification Traps. Larger muscle fragments detected with a stereoscope may be mistaken as a hair fragment, a feather barb fragment, or other similar looking fragments. Suspect muscle fragments detected with the stereoscope should be isolated and observed with the compound microscope.

A number of potential misidentification traps can be present when examining the sample with a compound microscope. Some mineral fragments may be misidentified as muscle fragments. Mineral fragment generally have an intense birefringence and therefore observing the suspect fragment with polarized light is recommended. Vascular bundles from plant fragments (e.g., helical xylem cells) presenting lignin deposits may have a similar appearance to the muscle striations, but the presence of a cell wall should help in the identification. In addition, due to the structural organization of the cellulose microfibrils in the plant cell walls, the majority of plant fragments will display birefringence under polarized light.

Bone

If present in high enough quantities, or if they are relatively large, bone fragments can easily be detected with a stereoscope. Bone fragments primarily concentrate in the sediment.

Fish bone fragments are readily distinguishable from mammalian and avian bone fragments. Differentiating mammalian and avian bone fragments is more challenging. In cases where only a few bone fragments are detected and it cannot be determined if the source is mammalian or avian; but it can be determined that the bones are not fish, the fragments can be identified as "terrestrial" bone fragments. This classification of "terrestrial" bone is commonly accepted and in use in the United States, Canada, and the EU.

The following stereoscopic and histological descriptions describe bone fragments that have principally not been stained with Alizarin Red. This staining procedure is nondestructive; therefore, the structural features of bone described later remain intact, but stained bone fragments are red to fuchsia in color.

Stereoscopic Features. Mammalian bone fragments have a dense chunky appearance under the stereoscope (Fig. 5.11) and can sometimes be mistaken as ground limestone (see Methods VII and XV in Chapter 9 Microchemical Spot Tests for differentiation tests). Usually the bone fragments are white in color, but high temperature treatments can make them range from white to light yellow to light tan to grey or even brownish. Two types of bone fragments can be distinguished: spongy bone (all epiphyses and the central medullar region of all long bones, vertebrae, skull bones, and sternum) and compact bone (restricted to the periphery of the long bone metaphyses). Spongy bone fragments, like the name suggests, have a sponge-like appearance composed of trabeculae. In comparison, compact bone fragments appear denser with numerous small oval pits or crevices punctuating their surface. Usually the spongy bone fragments are not present as frequently in mammalian bone as they are in avian bone, but this proportion depends on the type of bones being processed in the meat and bone meal.

Avian bone fragments are generally fragmented into splinter-shaped pieces with jagged edges (Fig. 5.12). The fragments are usually more yellow in color than mammalian bones but can range from white to light tan to yellow. Much more of the spongy bone fragments are present relative to bone from mammalian sources. The presence of eggshell fragments may also be present.

Fish bone fragments have a lustrous, often semitransparent appearance (Fig. 5.13).

Fig. 5.11. Bone fragments isolated from a mammalian meat and bone meal observed using a stereoscope. © Her Majesty the Queen in Right of Canada (2010).

Fig. 5.12. Bone fragments isolated from a poultry meal observed using a stereoscope. © Her Majesty the Queen in Right of Canada (2010).

Fig. 5.13. Bone fragments isolated from a fish meal observed using a stereoscope. © Her Majesty the Queen in Right of Canada (2010).

Fig. 5.14. Fish scales isolated from a fish meal observed using a stereoscope. Note: Most fish scale fragments are much smaller in size. © Her Majesty the Queen in Right of Canada (2010).

These bone fragments are more fragile than avian or mammalian bone fragments and can be broken relatively easily. Some fragments are very slender and cylindrical and may be mistaken as a feather shaft. Although present less frequently than the bone fragments, fish scales are normally detected among the fish bone fragments. The scales are generally translucent and flat glossy structures (Fig. 5.14). Scales often present step-like concentric growth rings that may be visible at a higher magnification (i.e., 30–50×).

Histological Features. Relative to the stereoscopic examination, much more structural details of bone fragments can be visualized using a compound microscope.

The lacunae (small holes of ~10–15 µm in the bone where the osteocytes, or bone cells, were located) and canaliculi (minute canals that radiate from the lacunae and carried nutrients to the osteocytes) structures are visible and can be used to help identify the origin of the bone (i.e., mammalian, avian, "terrestrial," or fish). Some bone fragments are large enough that the Haversian canals from the osteon (which surrounded blood vessels and nerves) are intact and visible. At the immediate vicinity of the Haversian canals, lacunae are organized in a circular pattern surrounding the canal.

To visualize the bone fragments, it is highly recommended that a clear resin that can be quickly fixed (e.g., Norland Optical Adhesive 65) be used to prepare a slide. This will ensure that the lacunae and canaliculi are not filled with the mounting medium as the trapped air in these structures increases their contrast relative to the bone thus making them easier to observe. Mounting media that remain fluid can be used to observe bone fragments (e.g., water, glycerol, paraffin oil, etc.) but will fill in these structures in time (i.e., <5 min) making them difficult, if not impossible, to distinguish. If such a medium is used to observe bone fragments, the slide must be observed immediately. If no fixed mounting medium can be used, then pure glycerol is highly recommended as it somewhat delays the fading of lacunae and canaliculi.

NOTE: Non-permanent slides can nevertheless always be observed later on by use of DIC, provided the refraction index of the mounting medium is high enough (glycerol n = 1.47).

When illuminated with polarized light, bone fragments may tend to have a moderate birefringence (in different color range if a

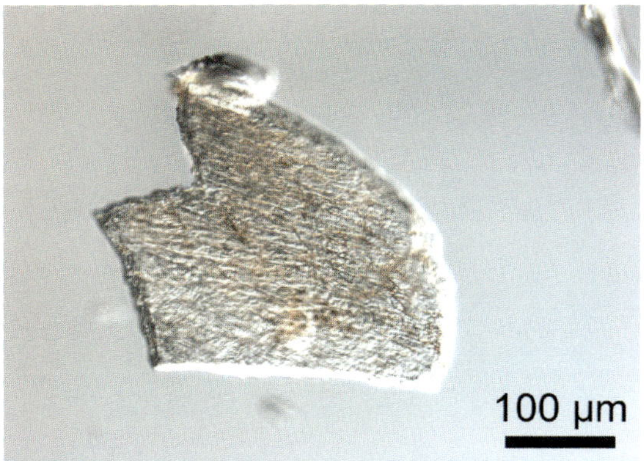

Fig. 5.15. Bone fragment isolated from a fish meal observed using a compound microscope and DIC lighting. © Her Majesty the Queen in Right of Canada (2010).

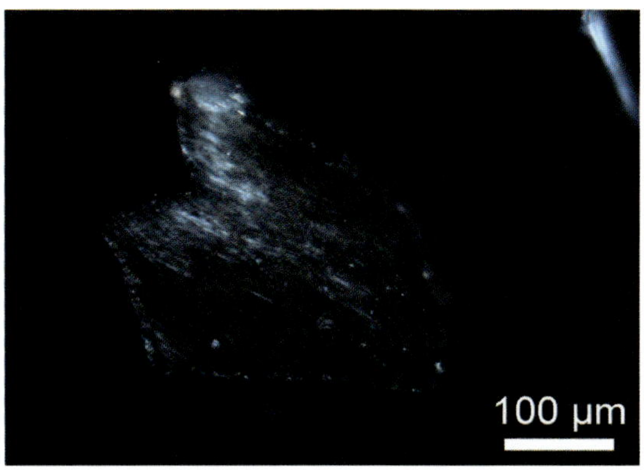

Fig. 5.16. The same bone fragment isolated from a fish meal as the previous figure observed using a compound microscope and polarized light. © Her Majesty the Queen in Right of Canada (2010).

$\lambda/4$ filter is used) (Figs. 5.15 and 5.16). This polarization of bone is due to the remains of collagen, the major component of the bone organic matrix. Old bones (e.g., removed from the soil) no longer have traces of collagen and thus are no longer birefringent compared to fresh bones. Heat treatment during meat and bone meal processing considerably destroys the collagen resulting largely in the loss of the birefringence.

Lacunae and canaliculi appear as black structures against a light gray, white, or semitranslucent background (or pink-red

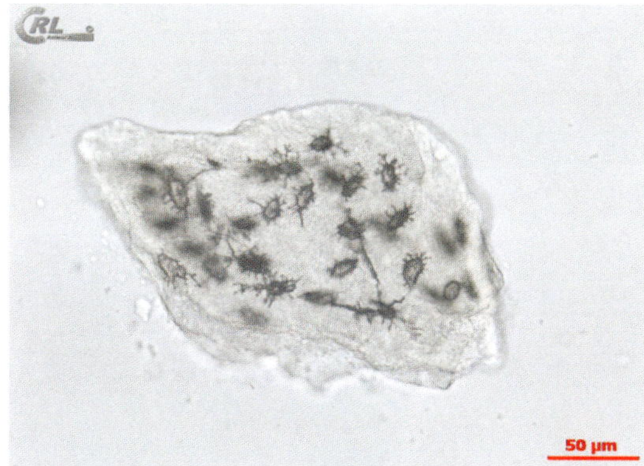

Fig. 5.17. Bone fragment isolated from a mammalian meat and bone meal observed using a compound microscope. The lacunae are globular. Micrograph by courtesy of CRL-AP.

background when stained with Alizarin Red). These structures are essential for identifying a small fragment such as bone. Careful observation of the lacunae, canaliculi, and other structural features of bone fragments can allow a microscopist to determine the origin of the bone fragment; i.e., mammalian, avian, "terrestrial," or fish.

Mammalian bone fragments tend to fragment as blunt, chunky pieces with few sharp edges (Fig. 5.17). The lacunae are generally elliptical to almost globular in shape (ratio from 1.2 to 2.5).

The number (i.e., density) of the lacunae of a bone fragment varies from very few to numerous according to the type of bone (spongy or compact). The pattern of distribution is irregular for spongy bone fragments and more organized in the compact bone area from the long bones. The network of canaliculi is fine and can be difficult to detect but are usually present and visible. Special note of the lacunae located close to Haversian canals. Some of them may appear as very elongated (ratio >2.5) in one direction with canaliculi radiating at a right angle making them resemble fish lacunae (Figs. 5.18 and 5.19). This is particularly true when they are observed on very small bone fragments that do not allow detecting the circular distribution pattern around the Haversian canal.

Fig. 5.18. Bone fragment isolated from a mammalian meat and bone meal observed using a compound microscope. The lacunae are elongated and the canaliculi are radiating at a right angle. © Her Majesty the Queen in Right of Canada (2010).

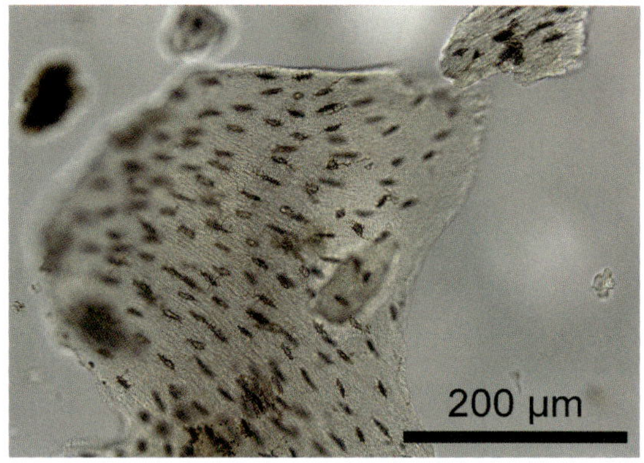

Fig. 5.20. Bone fragment isolated from a poultry meal observed using a compound microscope. The lacunae are flattened. © Her Majesty the Queen in Right of Canada (2010).

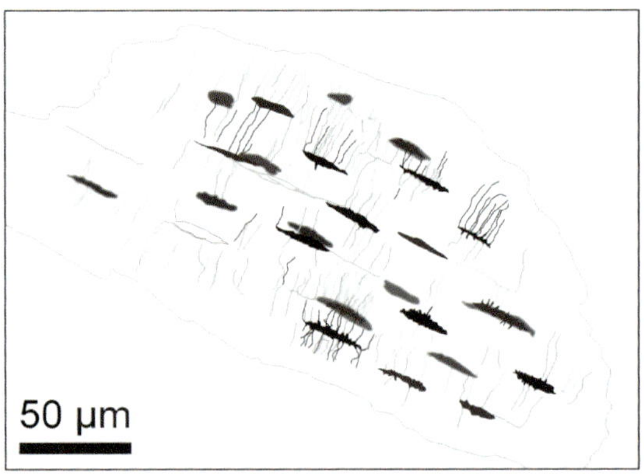

Fig. 5.19. Line drawing representation of Fig. 5.18 with the lacunae and canaliculi highlighted. © Kenneth Howe.

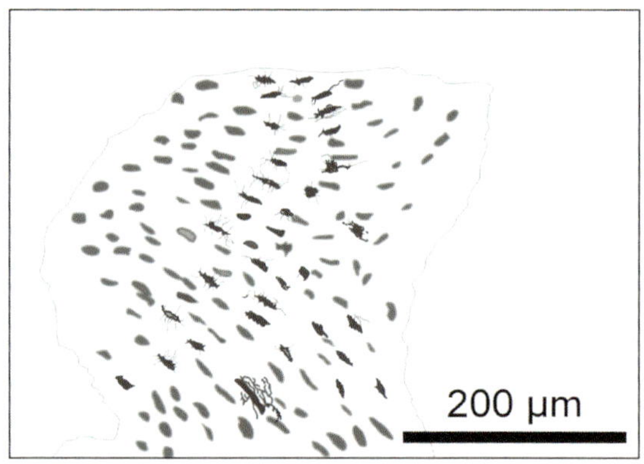

Fig. 5.21. Line drawing representation of Fig. 5.20 with the lacunae and canaliculi highlighted. © Kenneth Howe.

Avian bone fragments tend to have sharp edges. The lacunae range from flattened to oblong (ratio >2) (Figs. 5.20 and 5.21) or almost perfectly round in shape (ratio from 1.0 to 1.5) (Figs. 5.22 and 5.23). The density of the lacunae generally ranges from several to quite numerous (in general this density is higher than that of mammalian bones). No pattern of distribution can be assigned. The canaliculi of flattened to oblong shaped lacunae are generally present but faintly visible. The canaliculi of globular lacunae may not always be present; when present they tend to be relatively short in length and thicker as compared to those found in mammalian bone fragments.

Fish bone fragments present a large diversity of types and shapes. Fish bones can be classified in different structures:

- fish bones "*in se*"
- gills
- otoliths
- scales, and
- teeth.

In the present chapter only generalities on fish bones "*in se*" are described. Other bony fish structures are extensively described in Chapter 4 Feed Ingredients of Marine Origin.

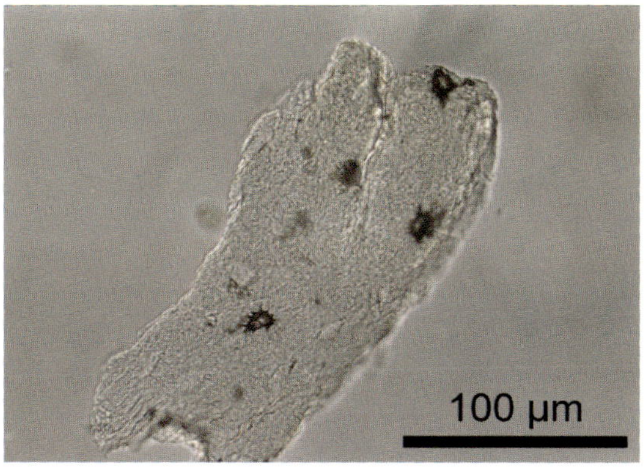

Fig. 5.22. Bone fragment isolated from a poultry meal observed using a compound microscope. The lacunae are round. © Her Majesty the Queen in Right of Canada (2010).

Fig. 5.24. Fish bone observed using a compound microscope. The lacunae and canaliculi have an appearance reminiscent of a neural cell. © Her Majesty the Queen in Right of Canada (2010).

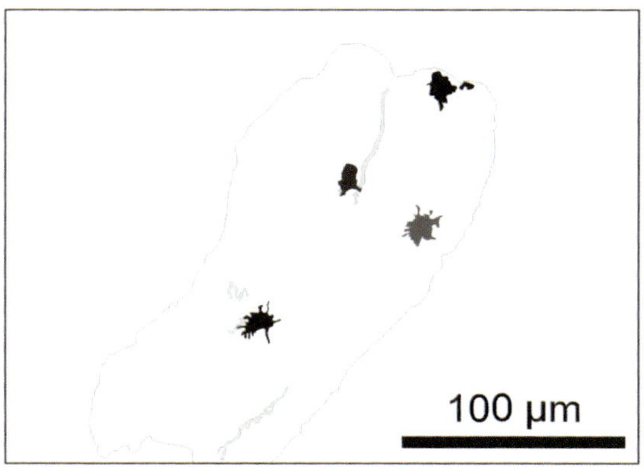

Fig. 5.23. Line drawing representation of Fig. 5.22 with the lacunae and canaliculi highlighted. © Kenneth Howe.

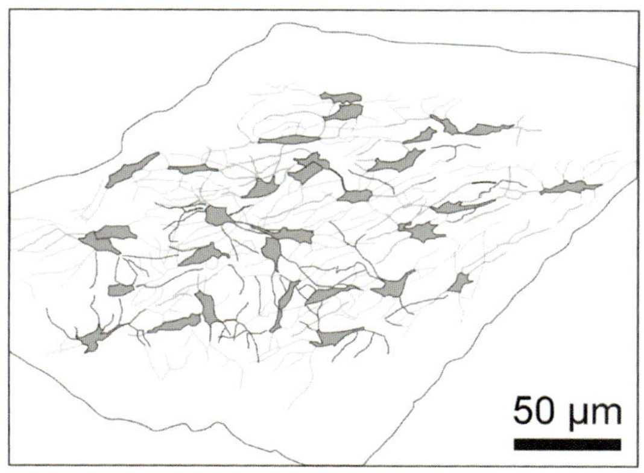

Fig. 5.25. Line drawing representation of Fig. 5.24. © Kenneth Howe.

Fish bones *"in se"* are generally easily distinguishable by the distinct appearance of their lacunae and canaliculi when they are visible—some fish bone particles sometimes lack lacunae. Generally their lacunae are flattened elongated to oblong (ratio >> 2.5) and the canaliculi are relatively thick and extend far from the lacunae usually emerging from them at right angles. Together the lacunae and canaliculi have an appearance reminiscent of a neural cell with projecting dendrites (e.g., herring bones) (Figs. 5.24 and 5.25). Alternately, lacunae in fish bones may be present as elongated parallel lines of varying thickness (Figs. 5.26 and 5.27) or as thick fissures with varying orientations (Figs. 5.28 and 5.29).

Aside from these descriptions, there exist many differences of lacunae shape and organizational patterns between fish bones depending on the type of bone formation and the ossification level. Various intermediates between bone and cartilage are commonly found in fishes. This explains why fish bone fragments present a great variability of staining intensities using Alizarin Red. Further detailed information can be found in Chapter 4 Feed Ingredients of Marine Origin.

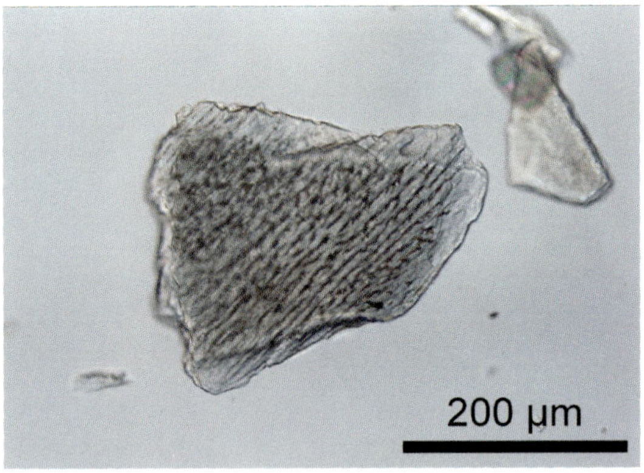

Fig. 5.26. Fish bone observed using a compound microscope. Note the elongated parallel lines of varying thickness. © Her Majesty the Queen in Right of Canada (2010).

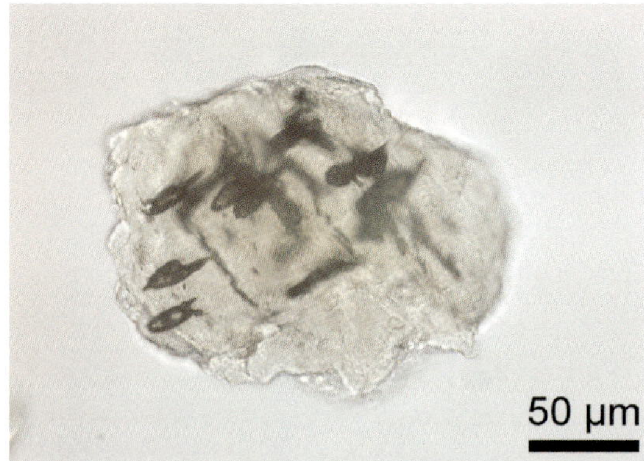

Fig. 5.28. Fish bone observed using a compound microscope. Note the thick fissures with varying orientation. © Her Majesty the Queen in Right of Canada (2010).

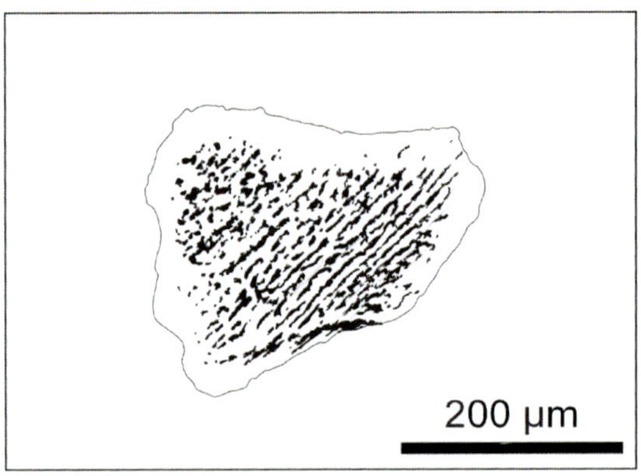

Fig. 5.27. Line drawing representation of Fig. 5.26. © Kenneth Howe.

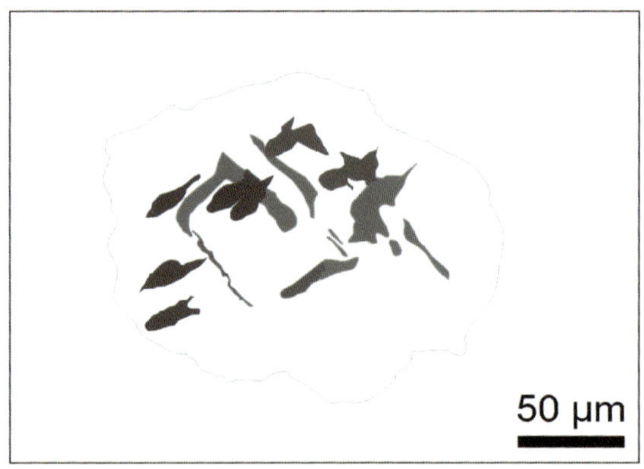

Fig. 5.29. Line drawing representation of Fig. 5.28. © Kenneth Howe.

Cartilage

Different types of cartilages exist: hyaline (e.g., in articulations), elastic (e.g., in ears), and fibrous (e.g., in intervertebral discs). All cartilage types are characterized by an extracellular matrix composed of collagen (different types) being secreted by chondrocytes or cartilage cells.

Cartilage fragments are present regularly in animal meals but in quantities that will vary with the type of meal and its animal origin. Only scarce amounts will be found in pure meat meal and few will be detected in mammalian meat and bone meals. In poultry meals this quantity will be higher but also depends on the age of the animals being processed (e.g., cartilage is more abundant in broiler chicken than in laying hen meals). In fish meals, cartilaginous structures are quite common, but their abundance is dependant from the type of fish species and the geographical origin of the fish.

Cartilage has a relative high density and concentrate in the sediment.

Stereoscopic Features. Particles of cartilage are difficult to identify with the stereoscope. They have a whitish to dirty yellow color and are opaque to semitransparent often with a fatty appearance. Their shape is highly variable, making them difficult to differentiate from other bone fragments or minerals present in the sediment. Examination with the compound microscope is highly recommended to confirm if cartilage is present.

Histological Features. Features of cartilage are similar for "terrestrial" animals (i.e., mammals and birds) and fish.

Cartilage fragments have a noticeably high density of lacunae (small holes where the chondrocytes, or cartilage cells, were located). The lacunae of cartilage differ from that of bone as they are exclusively spherical or slightly oval in shape, and their diameter is somewhat larger (~20–25 µm). In addition, lacunae in cartilage never have canaliculi. With very few exceptions, there is no pattern of lacunae distribution.

Fragments of cartilage seen with the compound microscope appear as more or less translucent, irregular, smooth edged particles. The abundant presence of spherical holes or pits disrupting a smooth surface, due to the high density of lacunae, is typical for cartilage surfaces or edges. Provided no severe heat treatment, they show a disorganized birefringence when observed by polarization. This birefringence is due to the abundance of collagen fibrils. Depending on the ossification level (endochondral bone formation within long bones epiphyses or ossification in fish bones—see fish bone "*in se*" description) and the concentration of hydroxyapatite, hyaline cartilage may be stained by Alizarin Red if mineralization occurred (Fig. 5.30). However, staining intensity is weaker than that observed of bones.

Feathers

The presence of feathers can be detected with both the stereoscope and the compound microscope. They can be present in a feed either raw or processed (i.e., hydrolyzed) form or both. Feather fragments primarily concentrate in the flotate.

Stereoscopic Features. Raw feathers tend to be easily detected by the presence of the large central shaft (i.e., rachis) of a feather with finer barbs branching off (Fig. 5.31). The length of the barbs branching off the central shaft is dependant on the physical stresses the feather has gone through, but the fragment is usually recognizable as a damaged feather. The rachis

Fig. 5.30. Cartilage fragment strained with Alizarin Red illustrating the high density of lacunae. Micrograph by courtesy of CRL-AP.

Fig. 5.31. Raw feather fragments observed in a poultry by-product meal using a stereoscope. The feathers are damaged, but the central shaft (rachis) is visible with the finer barbs branching off of it. © Her Majesty the Queen in Right of Canada (2010).

Fig. 5.32. Hydrolyzed feather meal observed using a stereoscope. © Her Majesty the Queen in Right of Canada (2010).

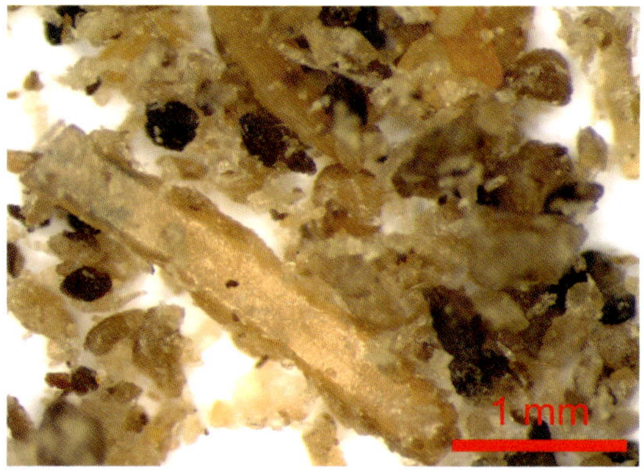

Fig. 5.33. Central shaft (rachis) of a feather fragment from a hydrolyzed feather meal observed using a stereoscope. The barbs are missing. © Her Majesty the Queen in Right of Canada (2010).

may be misidentified as a fish bone fragment or a hair fragment. Therefore, suspect fragments should be verified with the compound microscope. Barbules may be present and observed as very fine fibers that appear to have tiny black dots evenly spaced the entire length of the fibers when viewed at a relatively high magnification (i.e., 30× to 50×). These black dots are the nodes of the barbules that are more readily observable with the compound microscope. Rendered poultry meals and poultry by-product meals contain feathers that are generally processed to the point where only small feather fragments are detectable.

Hydrolyzed feathers can be more difficult to detect with the stereoscope due to the level of processing (Fig. 5.32). The central shaft of the feather is fragmented and the barbs are usually missing (Fig. 5.33), leading sometimes to the misidentification of the fragment as a fish bone fragment, hair fragment, or muscle tissue. Fragments appear as shiny cylinders that are round to oval in cross-section and vary in color from a translucent clear to caramel to brown and sometimes grey to black. Fragments are rarely larger than 5 mm in length and 1 mm in diameter and can be considerably smaller.

Histological Features. Raw feathers are readily recognizable by the presence of barbules. Barbules have a characteristic appearance of thin filaments with a series of evenly spaces

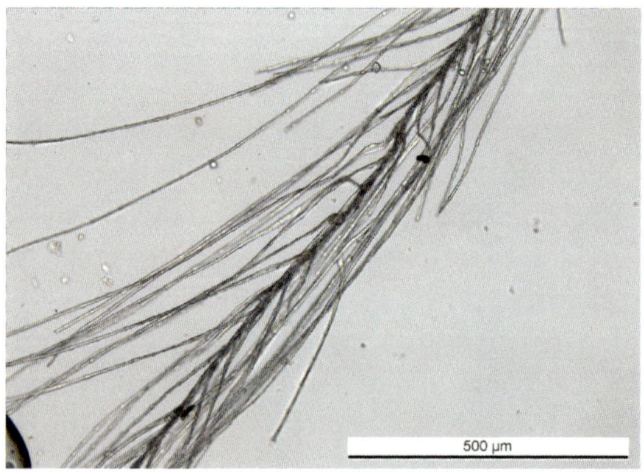

Fig. 5.34. Raw feather fragment observed using a compound microscope. The barbules are attached to the central barb. The nodes are visible as clear, evenly spaced bumps along the length of each barbule. The internodes are located between the nodes. © Her Majesty the Queen in Right of Canada 2010).

nodes. The barbules of raw feathers are generally fairly long and consist of a series of internodes and nodes. This alternation of nodes and internodes may resemble hyphae filaments from fungi. Often the presence of multiple barbules attached to a barb can be detected (Fig. 5.34).

Generally the barbs and barbules are translucent, but they can also be found with some pigmentation. Occasionally the presence of

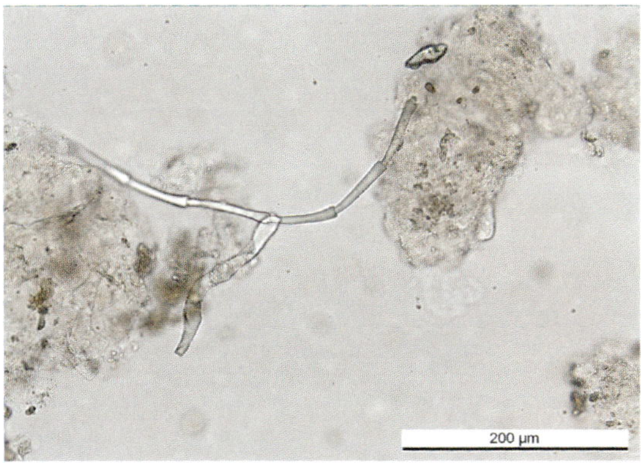

Fig. 5.35. Raw feather fragment observed using a compound microscope. This fragment consists of only one barbule with five nodes and six internodes clearly visible. © Her Majesty the Queen in Right of Canada (2010).

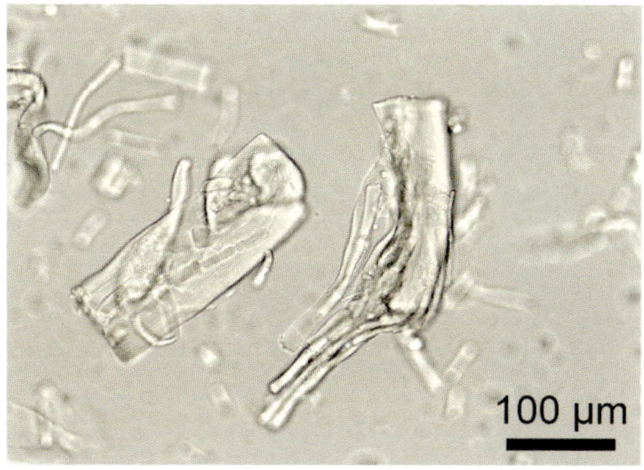

Fig. 5.37. Hydrolyzed feather fragments observed using a compound microscope. Two barbs are shown in this figure. The barbules appear to be melted around the barbs and are distinguishable by the nodes. © Her Majesty the Queen in Right of Canada (2010).

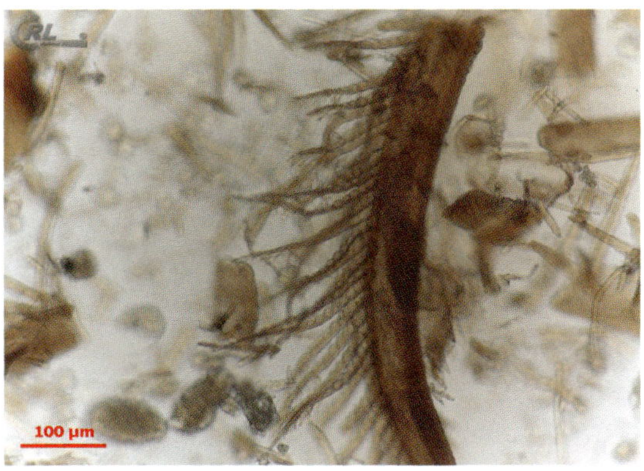

Fig. 5.36. Raw feather fragment stained by the cystine reagents turns brown. Micrograph by courtesy of CRL-AP.

barbules with barbicels (i.e., tiny hooks) can be detected, but the majority of barbules are simply a long series of nodes and internodes. Short feather fragments consisting of only a few nodes and internodes tend to be found in poultry meals and poultry by-product meals (Fig. 5.35). When observed under polarized light, some raw feather fragments may occasionally exhibit birefringence; however, the majority of them lack this feature. As feathers have high concentrations of cystine, a dimmer of the amino acid cysteine linked by a disulfide bridge, they react by staining dark brown when using the cysteine reagent (Fig. 5.36—as described in Chapter 9 Microchemical Spot Tests—Method XXII).

This allows differentiating easily some rachis fragment looking like fish bones or scales from real fish structures which never turn brown.

Hydrolyzed feathers can be more difficult to distinguish than raw feathers due to the structural degradation caused during processing. Barbs generally remain in a cylindrical shape, but the barbules have a melted and often conglomerated appearance. More often than not, the basal portions of the barbules remain attached to the barbs and appear to be melted around the barbs (Fig. 5.37). Unlike raw feathers, the barbules of hydrolyzed feathers are not relatively straight but rather twisted and tangled (several barbules may be intertwined and have the appearance of cooked spaghetti, Fig. 5.38). Although the barbules have been heavily damaged, the nodes of the barbules remain distinguishable. Short barbule fragments (e.g., consisting of one internode and one node) tend to appear melted and warped (Fig. 5.39). The processing of the feathers may give the barbs and barbules a yellow to orange to caramel color as opposed to the generally clear color of raw feathers. When observed under polarized light, hydrolyzed

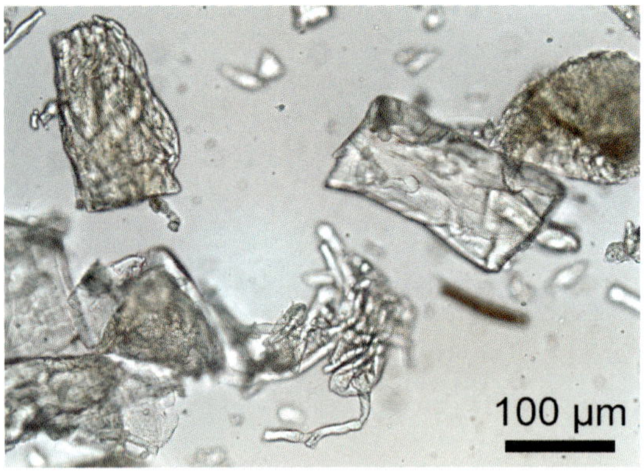

Fig. 5.38. Hydrolyzed feather meal observed using a compound microscope. A number of barbules are intertwined at the bottom center of the figure, giving it an appearance of cooked spaghetti. © Her Majesty the Queen in Right of Canada (2010).

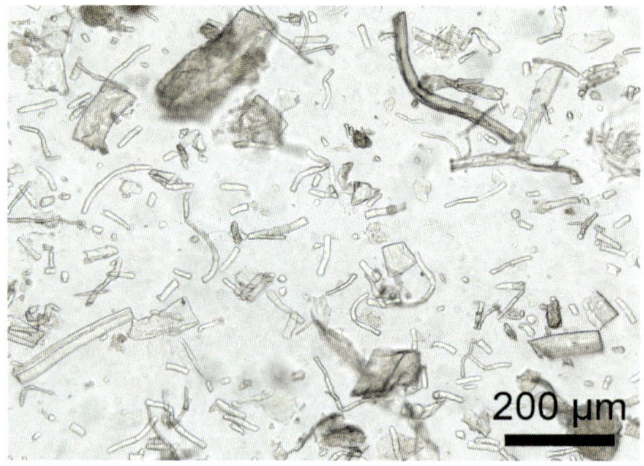

Fig. 5.39. Hydrolyzed feather meal observed using a compound microscope. Numerous short barbs and barbules having a warped appearance are present. © Her Majesty the Queen in Right of Canada (2010).

feathers may exhibit relatively little to no birefringence. Usual processing treatments do not reach a temperature above 150C upon which cystine disulfide bridges are broken, thus hydrolyzed feathers will also react with cystine reagent as raw feathers do.

Hair

Hair fragments tend to concentrate in the flotate and are usually detected with the stereoscope. For most species of interest to a feed microscopist, the hair fragments are too large to end up in the fine material used to prepare a slide for examination with the compound microscope. Hair fragments are therefore primarily detected with the stereoscope. Once detected, they should be removed from the light fraction, mounted in water, and examined with the compound microscope so that the histological details can be observed.

Hair will be found in a feed raw, rendered, or hydrolyzed. Raw hairs tend to be long and may include the basal (root) and/or distal (tip) end(s). Rendered hairs fragment into short segments (generally <5 mm). The shorter the hair fragment, the less distinguishable characteristics are present, thus making identification more challenging. As most hairs detected in feeds are rendered, this section will focus on characteristics that are useful in the identification of fragmented hair and ignore other characteristics that would be useful in the identification of full hairs (e.g., color banding patterns). Hydrolyzed hair may appear similar to rendered hair but depending on the level of processing may start to lose histological features. The cells of the medulla may no longer be recognizable and the cuticular scale may be damaged. Highly processed hydrolyzed hairs may no longer be recognizable as hair. This is typical for European meat and bone meals where higher temperature treatments are imposed by law. In these EU meals, hairs are no longer recognizable by microscopy. However, cases of hair detection in EU meals might originate from "natural" environmental contamination by rodents (e.g., rats, mice, etc.). The appearance of these detected hairs must therefore be documented as the taxonomic features allowing species identification can be used to support this manner of contamination.

It is highly recommended that microscopists obtain a reference collection of hair and fibers to allow for comparisons.

Stereoscopic Features. Raw hair appears as a long thin cylindrical fiber that gradually tapers

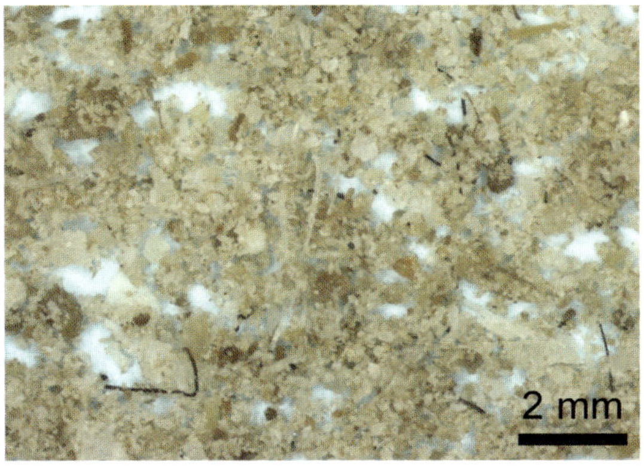

Fig. 5.40. Bovine meat and bone meal observed using a stereoscope. Several hair fragments are present as black, white, or caramel colored filaments. © Her Majesty the Queen in Right of Canada (2010).

Fig. 5.41. Hydrolyzed porcine hair observed using a stereoscope. The hair fragments are recognizable as hair. © Her Majesty the Queen in Right of Canada (2010).

to a point at the tip. The root may be present with follicular cells attached. Hairs are round, ovoid, or flattened in cross-section. Hair will vary in shades from white to grey to black and in color from blond to red to light brown to dark brown. The coloration may be consistent the entire length of the hair, or the hair may have a banding pattern.

Rendered hair is present as short cylindrical segments that are round, ovoid, or flattened in cross-section. Fragments are rarely larger than 5 mm in length and may be misidentified as a muscle fragment, a feather shaft or barb fragment, or a fish bone fragment. Generally, processed hairs are clear, white, caramel, brown, or black in color (Fig. 5.40).

Hydrolyzed hair may or may not be recognized, depending on the amount of hydrolysis. Hairs may look similar to rendered hair (Fig. 5.41) or can be processed to the point that it is no longer recognizable as a hair fragment (Fig. 5.42). Highly hydrolyzed hair may have a glassy appearance that is yellow, caramel, brown, or black in color. Highly processed hydrolyzed hair may be mistaken as a blood meal.

Histological Features. Hairs are composed of three distinct layers. Starting from the outside these layers are the cuticle, followed by the cortex, and then the medulla. Careful obser-

Fig. 5.42. Hydrolyzed porcine hair observed using a stereoscope. This sample has been processed to the point where the hair fragments are no longer recognizable as hair. © Her Majesty the Queen in Right of Canada (2010).

vation of all three layers of the hairs can allow the microscopist to identify the species of origin, a group of potential species, or to exclude species.

The vast majority of the histological features of hair fragments remain present even if the hair has been through the rendering process. The major structural damage sustained by hair during the rendering process is fragmentation along its length (i.e., fragmented into small cylindrical pieces). It has occasionally been observed that part of the scale may be damaged and missing from a hair fragment, but this is a rarity and is detected primarily with porcine bristles.

The histological features of hydrolyzed hair, on the other hand, may not be present, depending on the level of processing. Hydrolyzed hairs can appear similar to that of rendered hair or may be processed to the point that it is no longer recognizable as a hair. Generally, raw hair and rendered hair fragments exhibit birefringence when observed with polarized light while hydrolyzed hairs exhibit relatively little to no birefringence.

A few general observations of the hair fragment should be noted when examining it, including the diameter, the cross-section, and the general shape of a hair. The diameter can be quite variable between species, the type of hair (e.g., guard hair vs. undercoat hair), and the location on a single hair (i.e., basal vs. distal area). Fortunately, hairs usually have a diameter range in which they fall, although the presence of exceptions should be noted. The cross-section can be observed, as some species tend to have more rounded hairs while others tend to be more ovoid to flatten. If a hair fragment is sufficiently long, the overall shape can also help in the identification. Some hair fragment will appear to be relatively straight while others may be wavy, curly, or twisted.

The cuticle is composed of the outer scales of the hair and is found in a variety of configurations between various mammalian species. Scales always develop such that the base of the scale is located towards the basal end of the hair while the outer margin of the scale is closer to the distal end. A scale may be large enough that it will encompass the entire diameter of the hair, or they may be small enough that several scales are required to circle the diameter of the hair. The outer margin of the scale creates an observable scale pattern that can range from regular to irregular and may or may not overlap one another. The scale margin spacing ranges from close to distant and the texture of the margin can range from smooth to ripple to crenate. Examples of different scale patterns and scale margins are shown in Figs. 5.43, 5.44, 5.47, 5.48, 5.51, 5.52, and 5.54. Often the cuticles of hydrolyzed hair fragments have a number of fissures present throughout. The scale pattern may be visible upon close examination of the hair.

In some instances the hair fragment has such dark coloration that the scale pattern is not observable. In other instances the optics of the compound microscope may not be of high enough quality to observe the fine details of the scale pattern. In either such cases, an impression of a hair can be cast and then

Fig. 5.43. Bovine hair fragment observed using a compound microscope and DIC light. The scale pattern is irregular wave mosaic. © Her Majesty the Queen in Right of Canada (2010).

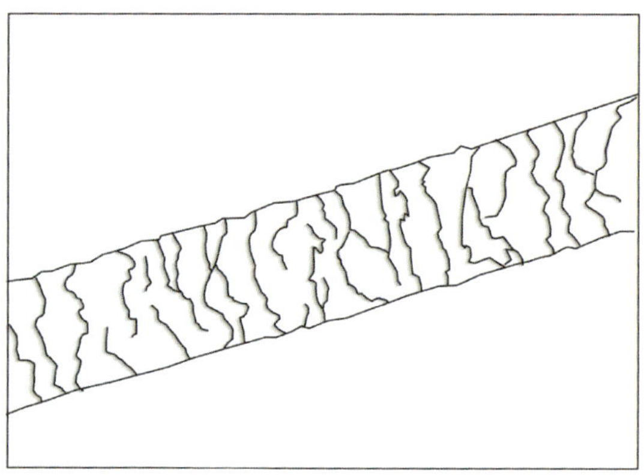

Fig. 5.44. Line drawing representation of Fig. 5.43. © Kenneth Howe.

observed with the compound microscope (Figs. 5.47, 5.48, and 5.54). To cast an impression, brush a thin layer of clear nail polish (Alternately, use a 1:1 mixture of clear nail polish and acetone. This mixture will be less viscous than nail polish and dry quicker.) onto a microscope slide, allow it to begin to dry for a few seconds, and then gently place the hair fragment on top. It may be necessary to gently puff on the hair fragment with one's breath to set it into the nail polish. Allow the nail polish to begin to harden and gently remove the hair, leaving the cast impression behind. Observe the scale pattern with the compound microscope.

An alternative to casting an impression of dark hairs is to remove some of the coloration. This can be accomplished by placing a hair fragment in a 3% (v/v) aqueous hydrogen peroxide solution and then warming the solution to 60C for 1 h. This procedure is more labor intensive than casting an impression but offers a few key advantages. First, the intensity of the hair color is muted thus allowing more light to pass through thus allowing all of the histological features of the hair to be more readily observed, including the internal structural features described below. Second, the hair pigments are not removed from the hair thus the pigment configuration can be observed. Third, small hair fragments are difficult to manipulate when attempting to cast an impression of the scale.

The cortex is the next layer in the hair. The thickness of the cortex is generally measured in relation to the thickness of the medulla (see later). The cortex is where the majority of the hair coloration is located. The hair may have a stained appearance (i.e., color is present but color pigments are not distinguishable),

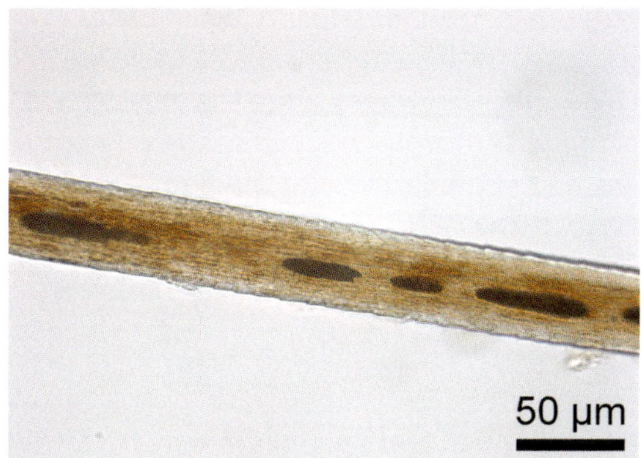

Fig. 5.46. Bovine hair fragment observed with the compound microscope. The medulla is broken. © Her Majesty the Queen in Right of Canada (2010).

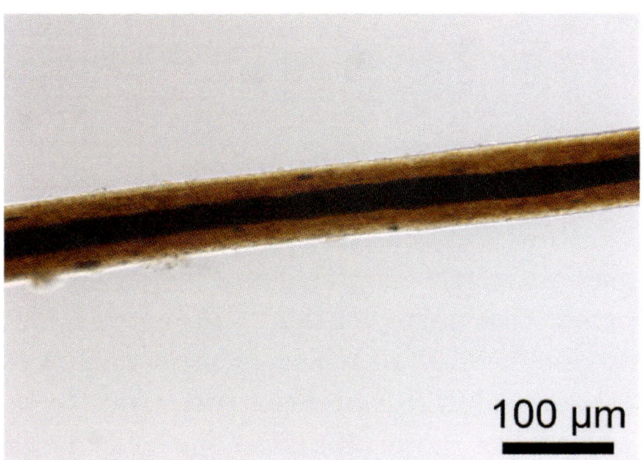

Fig. 5.45. Bovine hair fragment observed using a compound microscope. The pigments are primarily present as strands; a few ovoid pigments are also present near the periphery. The medulla is continuous. © Her Majesty the Queen in Right of Canada (2010).

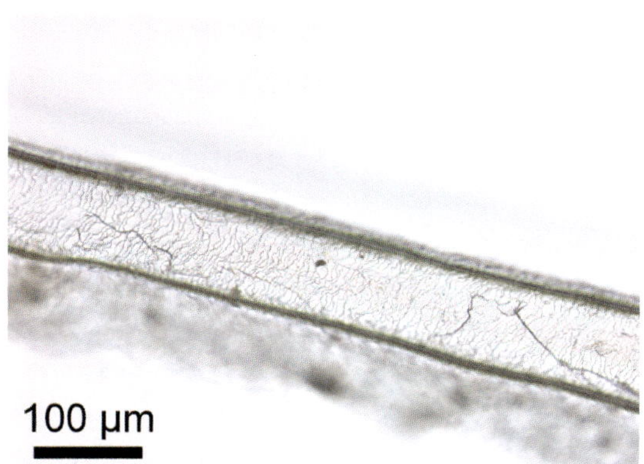

Fig. 5.47. A cast impression of a porcine bristle observed with a compound microscope. © Her Majesty the Queen in Right of Canada (2010).

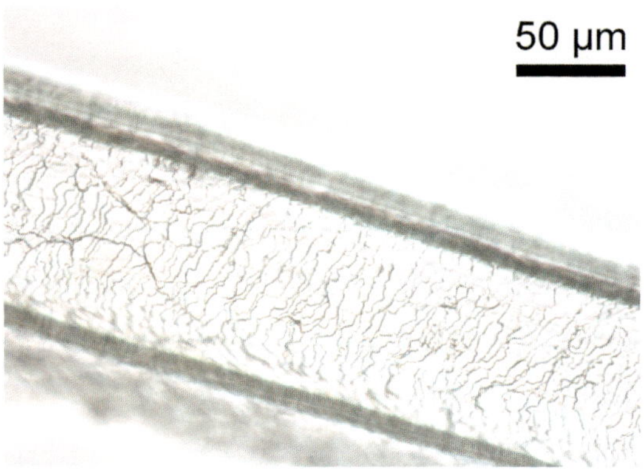

Fig. 5.48. A cast impression of a porcine hair bristle observed using a compound microscope. The margins of the scale are close, irregular waves. Note the large cracks running somewhat parallel to the edge of the hair at the left side of the figure. © Her Majesty the Queen in Right of Canada (2010).

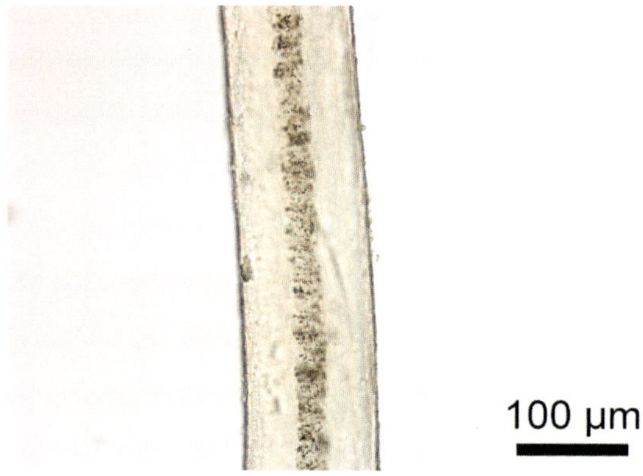

Fig. 5.49. Porcine hair fragment (bristle) observed using a compound microscope. The medulla running along the center of the hair fragment is clearly visible. © Her Majesty the Queen in Right of Canada (2010).

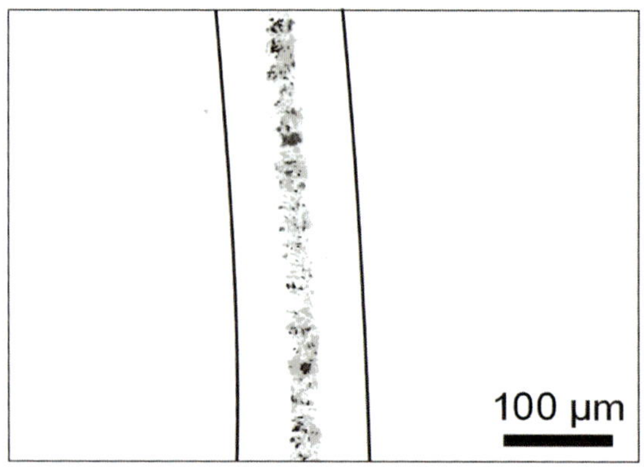

Fig. 5.50. Line drawing representation of Fig. 5.49. © Kenneth Howe.

or individual color pigments may be found (or a combination of both). Pigments have a variety of configurations including granules, strands, clusters, and round to ovoid to flattened bodies. White and grey hairs lack coloration.

The medulla is located in the centre of the hair. Many configurations of the medulla are possible. The medulla can be continuous throughout the length of a hair, or it can change. In several species, the medulla is continuous towards the basal end of a hair, then becomes fragmented and eventually absent towards the distal end. Most hair fragments found in feeds are short, and generally only one type of medulla pattern is discernable in individual fragments. Examples of different patterns are shown in Figs. 5.45, 5.46, 5.49, 5.50, 5.53, 5.55, and 5.56. Depending on the level of processing, the cells of the medulla may not be present in hydrolyzed hair fragments.

The histological features of several key species of interest to the feed microscopist have been included in tabular form. See Table 5.A.

Raw and rendered hair fragments display moderate birefringence when illuminated with polarized light. Hydrolyzed hair fragments may also display moderate birefringence, but this birefringence tends to be weaker the more processed the hydrolyzed hair is. Trichomes from various plants that may be mistaken as hair fragments tend to display an intense birefringence when illuminated with polarized light. Therefore, illuminating suspect hair fragments with polarized light is recommended to help identify the source of the fiber.

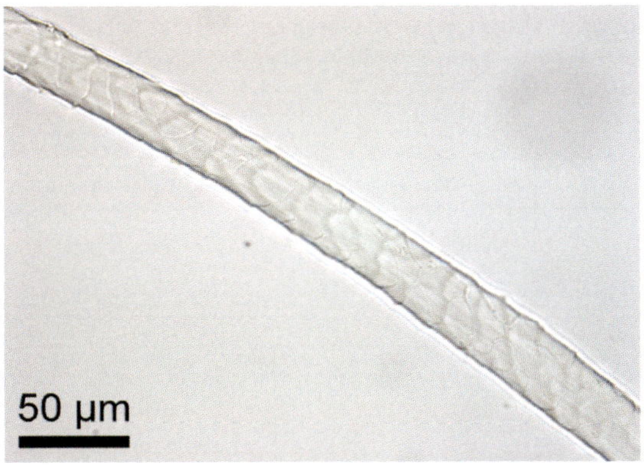

Fig. 5.51. Ovine hair fragment observed using a compound microscope. The scale pattern is distant to intermediate, smooth, irregular, petal to wave, mosaic. © Her Majesty the Queen in Right of Canada (2010).

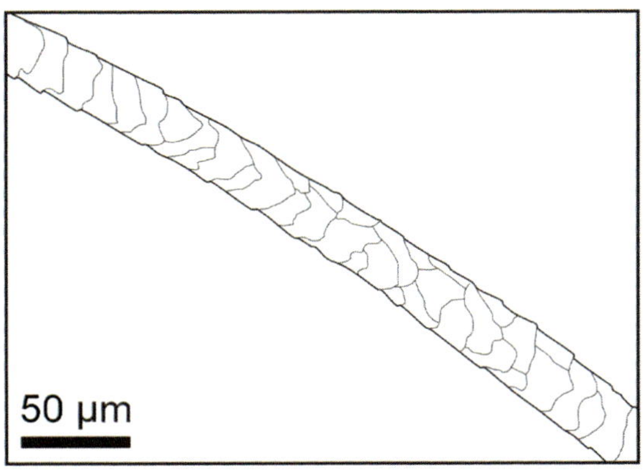

Fig. 5.52. Line drawing representation of Fig. 5.51. © Kenneth Howe.

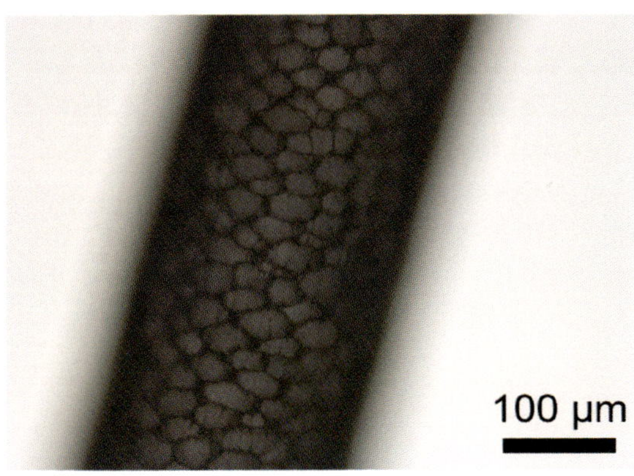

Fig. 5.53. Deer hair fragment observed using a compound microscope. The medulla is virtually the same diameter as the hair. © Her Majesty the Queen in Right of Canada (2010).

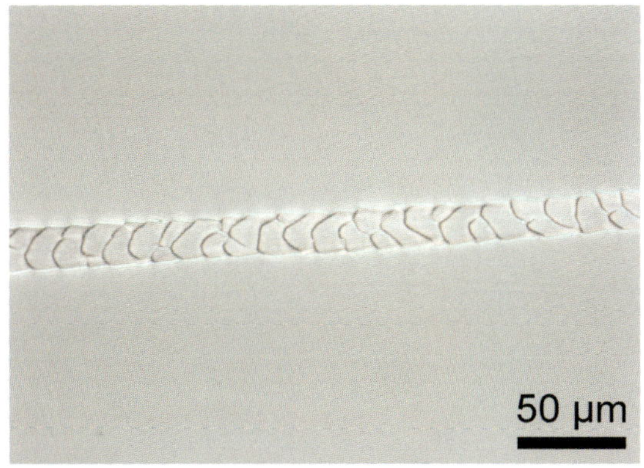

Fig. 5.54. A cast impression of a mouse hair observed using a compound microscope. The margins of the scale protrude noticeably and are distant, irregular, petal to wave, mosaic. © Her Majesty the Queen in Right of Canada (2010).

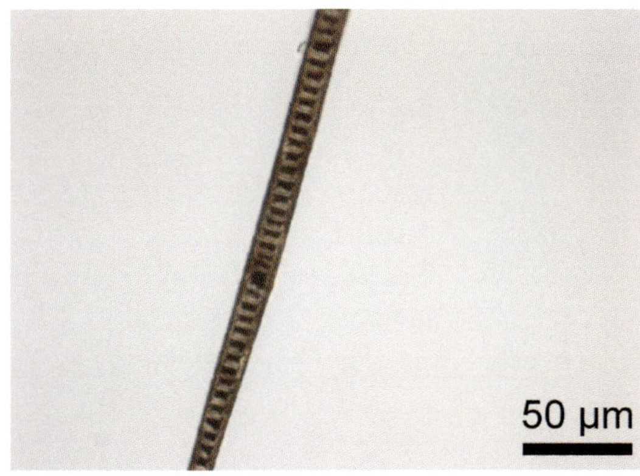

Fig. 5.55. Mouse hair fragment observed using a compound microscope. The pigmentation is stained. The medulla is a single ladder. © Her Majesty the Queen in Right of Canada (2010).

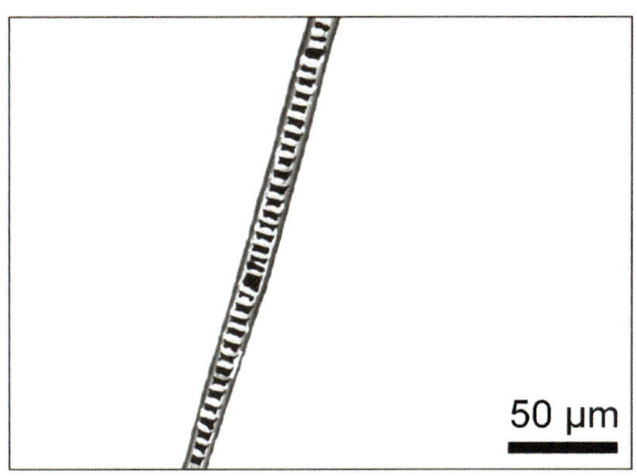

Fig. 5.56. Line drawing representation of Fig. 5.55. © Kenneth Howe.

Table 5.A. General Descriptions of the Histological Features of Various Hairs.

Histological Feature	Human	Cow/ Horse	Pig	Sheep	Deer/Elk	Rodent/ Feline
Diameter	Variable depending on race. Average of 90–100 µm, but can range from 15 µm to 180 µm.	50–60 µm, but can be as large as 130 µm.	90–180 µm. The diameter undulates along the length.	25–50 µm.	Up to 360 µm.	20–45 µm.
Cross-section	Round to ovoid to flattened.	Ovoid.	Round to ovoid.	Ovoid to flatten.	Round to ovoid.	Round to ovoid.
Scale Pattern	Basal: irregular wave mosaic. Tip: irregular wave.	Basal: irregular wave mosaic. Distal. irregular wave (Figs. 5.43, 5.44).	Irregular wave (Figs. 5.47, 5.48).	Smooth irregular petal to wave mosaic. Scales are often the diameter of the hair (Figs. 5.51, 5.52).	Basal: regular mosaic. Tip: regular to irregular mosaic to irregular wave mosaic.	Irregular petal to irregular wave mosaic (Fig. 5.54)
Scale Margin	Basal: close to intermediate; smooth with some serration. Tip: intermediate to distant; smooth. May appear frayed due to wear and treatments.	Basal: intermediate to distant; smooth. Distal. intermediate; crenate (Figs. 5.43, 5.44).	Close; smooth to rippled. Rendered hair fragments may appear to have large cracks (Figs. 5.47, 5.48).	Protrudes noticeably. Basal: distant Distal. distant to intermediate (Figs. 5.51, 5.52).	Basal: distant to intermediate; smooth. Distal. close to distant; rippled, smooth.	Protrudes noticeably. Smooth and distant (Fig. 5.54).

Histological Feature	Human	Cow/Horse	Pig	Sheep	Deer/Elk	Rodent/Feline
Pigmentation	May be absent (white hair). Stained, strands, granules, clusters, and/or ovoid to elliptical bodies.	May be absent (white hair). Stained, granules, clusters, strands, and/or ovoid bodies (Fig. 5.45).	May be absent (white hair). Fine granules or stained appearance.	Often absent (white hair). Stained or granules (Fig. 5.51).	Stained (Fig. 5.53).	May be absent (white hair). Large dark rounded to ovoid granules to stained (Fig. 5.55).
Medulla	Varies from absent to fragmental to unbroken. Amorphous, occasionally vacuated. Generally less than one-third the diameter.	Varies from absent to fragmental to unbroken. Amorphous, occasionally vacuated. Generally less than half the diameter (Figs. 5.45, 5.46)	Absent to continuous. Often very narrow and appear like fine granules. May be greater than one-third the diameter (Figs. 5.49, 5.50).	Absent (Fig. 5.51).	Unbroken lattice (Fig. 5.53).	Single or multiple ladder. (Figs. 5.55, 5.56).

Interpretation of Observations and Limitations of Microscopy

The following describes some general guidelines to interpreting observations of animal products in feeds. The information that can be obtained when identifying individual components of animal products is discussed in the following sections.

Blood

The presence of blood only reveals the presence of animal products. Usually blood indicates the presence of blood meal. Blood meals from all species are permitted in ruminant feeds in the United States and Canada. This, however, is not true in the EU. Blood products are not authorized in ruminant feeds but, provided they do not come from a ruminant, they are permitted in nonruminant feeds. However, many meat meals and meat and bone meals may have small amounts of blood present. If other animal markers are detected (i.e., bone fragments, muscle tissue, etc.), blood may not be present as an independent ingredient in the sample but as part of another potentially prohibited animal product. This is especially true if only a few pieces of blood are detected along with a relatively large number of other animal markers. The identification of blood meal is important especially if other methods (e.g., DNA based) are used to test feeds as it is possible that prohibited species are identified by these methods when the ingredient itself is exempt from the feed bans. Determining the species of origin of a blood product is not possible by light microscopy.

Milk Products

The presence of milk products simply indicates that milk products are present in the feed. Milk products from all mammalian species are permitted in ruminant feeds in the United States, Canada, and the EU. The identification of milk products is important, especially if other methods (e.g., DNA-based) are used to test feeds as it is possible that prohibited species are identified by these methods when the ingredient itself is exempt from the feed bans. Determining the species of origin of a milk product is not possible by light microscopy.

Muscle Tissue

The presence of muscle tissue indicates the presence of animal products. If muscle tissue is the only animal product detected in the sample, it may indicate the presence of a meat meal, meat and bone meal, poultry meal, poultry by-product meal, fish meal, etc. Although most of these products contain bone fragments that are usually more readily detectable than the muscle tissue.

Bones

The presence of bone fragments indicates the presence of animal products and can potentially allow the microscopist to classify the origin of the animal product. It is relatively simply to determine if a bone fragment is from a fish or "terrestrial" (i.e., mammalian or avian) origin. It is more challenging to determine if a terrestrial bone is from a mammalian or avian origin as there is much overlap in the histological features of these two classes. This can be simpler if several of the bone fragments are large enough to be observed with the stereoscope; but in many cases where a feed is contaminated with an animal product, it is common to only detect bone fragments with the compound microscope. It is recommended that numerous bone fragments be scrutinized before identifying a terrestrial bone as either mammalian or avian. A basic prudential rule would be to consider the observation of at least five particles of a given type of bone before making a decision on its origin, even for ascertaining the effective presence of contaminating bones for avoiding the risk of false-positive results.

Feathers

The presence of feathers indicates the presence of avian products in a feed. Raw feathers can be attributed to environmental contamination; but may also be present as a poultry meal or poultry by-product meal. If raw feather fragments are detected and other tissues are not (e.g., muscle tissue and avian bone fragments), environmental contamination has to be suspected but confirmation should be delivered by other methods such as DNA analysis (e.g., by using dedicated markers for pigeons or sea gulls). The presence of hydrolyzed feathers is indicative of the presence of hydrolyzed feather meal or poultry by-product meal.

Hair

Hair present in a feed generally indicates the presence of mammalian products and can lead to the identification of the species of origin of the product. However, raw hairs alone in a feed may indicate environmental contamination as opposed to contamination with an animal product. Microscopists should take the proper precautions to avoid contaminating a sample with one's own hair and/or the hair of a companion animal (e.g., cat, dog, etc.) that could be present on their clothing and introduced into the sample when preparing or examining it. Care should also be taken when a hair has an unnatural color, as this may be another indicator of environmental contamination. For example, it is not uncommon to find sheep hair (i.e., wool) that has been dyed an unnatural color (e.g., red, blue, etc.). Such a hair fragment is likely from an article of clothing or other textile. Additionally, care should be taken to avoid misidentifying plant trichomes and other fibers as hair.

The presence of processed hair generally indicates that the feed has been contaminated with a mammalian animal product. However, environmental contamination may be determined depending on the species identification of the hair. For example, rodent hair is distinct as compared to most species of interest to a feed microscopist (e.g., cow, pig, sheep, goat, deer, elk, etc.) and, if found to be processed, it may indicate that a feed ingredient was contaminated or had been in contact with rodents before processing.

The histological features of hairs that are commonly found in feeds are presented in Table 5.A.

Multiple Animal Markers

The detection of multiple animal markers in a feed will help a microscopist determine the animal product(s) present and the class or species of origin. One should be familiar with the different tissues that are present in various ingredients of animal origin (see Chapter 3 Feed Ingredients of Animal Origin and Chapter 4 Feed Ingredients of Marine Origin). The observations provided by a microscopist can help determine the most probable source of contamination. This information should be used in conjunction with the knowledge of potential sources of contamination which the feed could have been exposed to with a trace-back investigation from the sources of the individual feed components through the transportation of the ingredients to the production lines used to manufacture the feeds to the end use of the feeds (e.g., the feedlots or farm).

6 Fertilizer Microscopy

Value and Application of Fertilizer Microscopy

The value of the microscope in a scheme of analysis of fertilizers lies mainly as a supplement to chemical analyses. It can be used effectively for routine qualitative identification of organic and inorganic sources of the primary and secondary plant food elements (i.e., the actual forms of these which are present in the fertilizer mixture). Approximate quantitative proportions can be ascertained; completing and confirming the partial story told by chemical analyses.

Readers should have some basic knowledge of mixed fertilizers and materials, and a smattering of chemical terminology. Microscopic analysis of fertilizers brings to light discrepancies; this information can benefit both consumer and manufacturer. The mistake is often unintentional, possibly due to contamination of ingredients or human error.

Economic and Quality Perspectives

We quite often find discrepancies that are economically beneficial to the consumer and detrimental to the manufacturer, considering the relative commercial values of the product involved. In these cases, while the consumer can come out ahead as far as the commercial value of the fertilizer is concerned, this may be more than offset by detrimental effects of an undesirable material on a particular crop. The reverse can also occur, and the consumer gets less then he or she pays for, if the manufacturer omits or substitutes an ingredient, intentional or otherwise. Even a limited chemical and microscopic laboratory more than pays for itself by obviating such mistakes in manufacture. The state control laboratory will have a more effective means of protecting industry and consumer through microscopy.

By way of illustration, the need for fertilizer microscopy became particularly apparent several years ago in Florida in shade tobacco fertilizers. Several shade tobacco growers believed there was a substitution of ingredients in their fertilizers which adversely affected their crops. They were paying a premium price for a particular organic ingredient, which they believed essential to the quality of shade tobacco.

Another cheaper ingredient could be substituted which would have the same general appearance to the average person; it would also analyze the same, chemically. With the microscope, the growers' suspicions were confirmed. As a result of these findings, several lawsuits in the hundreds of thousands of dollars were instituted and successfully prosecuted.

Practical Application—Purity of Nitrogen Source

This hypothetical situation will show the practical value of fertilizer microscopy to a state control laboratory. Suppose chemical analyses determine that there is an excess of ammoniacal nitrogen present above the manufacturer's guarantee. The manufacturer has indicated in their guarantee 5.00% total nitrogen, of which 2.00% is ammoniacal, 2.00% is nitrate, and 1.00% is water-insoluble nitrogen. Ingredients such as ammonium nitrate, castor pomace, and sewage sludge (plus, of course, sources of phosphorus, potassium, etc.,—but we are concerned in this instance with only the nitrogen) are listed. However, chemical analysis indicates 6.20% total nitrogen to be present, of which 3.05% is ammoniacal, 2.05% is nitrate, and 1.10% is water-insoluble nitrogen. The castor pomace and sludge would account for the water-insoluble portion, which is close to the guaranteed amount. But the 3:2 ratio of ammoniacal:nitrate cannot be accounted for by ammonium nitrate. This compound could be present, but if so, necessarily along with another source of ammoniacal nitrogen.

Microscopic examination with a stereomicroscope under low power (10–40×) confirms the presence of castor pomace and sewage sludge. In addition, spherical white pellets are observed. Chemical spot testing produces a positive test for nitrate and ammoniacal nitrogen. This confirms the presence of ammonium nitrate. Also observed are fairly large, oblong, hexagonal, white crystals, typical of several sulfates. It is noted that the manufacturer lists no sulfates among their ingredients. (Examples might be sulfate of potash, Epsom salts, manganese sulfate, etc.). On the assumption that this may be ammonium sulfate, a chemical spot test for ammonium and sulfate ions yields positive indications for both. From the foregoing, we can conclude that the manufacturer has put in ammonium sulfate in addition to ammonium nitrate as a nitrogen source, which is contrary to their guarantee.

Practical Application—Excess Chlorine

As another example, suppose a sample guarantees as follows: water-soluble potash, minimum 8.00%; chlorine, maximum 1.00%; water-soluble magnesia, minimum 2.00%. The ingredients listed pertaining to these elements are: sulfate of potash and "Sul-Po-Mag" (a double sulfate of potassium and magnesium). No chlorides are listed, which indicates that what little chlorine might be found would be from impurities. Chemical analysis reveals 8.25% water-soluble potash, 1.90% water-soluble magnesia, and 4.80% chlorine. In this case the potash and magnesia are as would be expected, but the chlorine is above the maximum tolerance permitted.

Excess chlorine may be harmful to crops, particularly tobacco, but it is an essential component of the cheaper of the two common sources of potash: potassium chloride. Making a logical assumption, one can estimate that about 2.00% out of the 8.25% potash found was from the "Sup-Po-Mag" (since domestic "Sul-Po-Mag" contains almost equal amounts of potash and magnesia expressed as oxides, and we found close to 2% magnesium oxide). This leaves us 6.25% potash to account for from another source. We know that approximately 75% of the potash (as K_2O) in a fertilizer formulated from potassium chloride will equal the chlorine roughly equal to the amount of chlorine we found: 4.80%. Under the microscope, the "Sul-Po-Mag" particles can quickly be observed due to their crystalline appearance. Chemical spot tests confirm the presence of magnesium, potassium, and sulfate ions in these particles. Next, the pink-splotched cubic crystals, which we suspect are potassium chloride, are removed. Spot tests identify the cation as potassium and the anion as chloride. Through further examination, we find no potassium sulfate crystals. The foregoing confirms our suspicions and gives us a basis for appropriate action. Without the supplemental microscopic identifications, chemical analyses would be inconclusive.

Value to Manufacturers

Manufacturers have found routine microscopic analyses of particular value in their own quality control programs. Costly or damaging errors in formulation can be discovered through routine examinations. One may keep a constant check on their sources of materials and readily discover contamination, adulteration, or substation of tankages and meals before they are used in a mixture, thereby eliminating a possible problem in the final product. By means of a simple flotation procedure (see Chapter 1, Methods of Agricultural Microscopy), a rapid quantitative estimation may be made, ascertaining whether a soybean meal is 44% or 50% protein. Qualitative spot tests quickly disclose the composition of an inorganic material. Through the discovery of a single instance of irregularity, a manufacturer can save several times the cost of establishing and maintaining an inspection program.

Several specialized fields are useful to the microscopist. Making use of a property of crystalline particles, their refractive indices are of value in confirming the identity of a crystal. This property, a physical function involving the bending of light rays at a measurable angle peculiar to a given crystalline compound, provides a means of positive identification of a fertilizer ingredient. Of course, this is restricted necessarily to only pure crystals that can be removed from the fertilizer mixture.

Photomicrography has a definite place in the microscopists' routine. Color prints are useful as references for standard materials and various mixtures. Once the technique has been developed through experience, photomicrographs can be made rapidly and can prove of inestimable value in referring back to questionable samples.

Technique and Equipment

As with any type of analysis, the value of microscopy depends completely on the sample furnished. If the sample is not uniform and representative of the lot of fertilizer from which it is drawn, then any subsequent analyses, chemical, microscopic, or other, are of no value. The preparation of the sample for low power microscopic work, however, differs from that usually followed before chemical analyses. While in chemical analysis, the sample is generally ground and sieved, the sample for low power microscopic examination should merely be well mixed to insure uniformity. The individual particles can be much more easily identified in their original size and shape.

Occasionally, samples must be examined under high power (100×) using a compound microscope. When compound microscopy is used, the sample must be placed on a glass slide and be thin enough for light to transmit through it. Thus, cellular structure and features can be ascertained to make an even more complete identification. This chapter will, however, be concerned primarily with the use of the stereomicroscope at low power with reflected light; and discussion will pertain to this unless otherwise specified.

Fertilizer microscopy is similar in many respects to the practice with feeds ingredients, so basic essentials and manipulation techniques are similar to those outlined in Chapter 1 Methods of Agricultural Microscopy. The system of scanning a sample and manipulation will vary with the analyst and develop with experience. Equipment need not be elaborate; the need or desirability of a particular item will depend upon the analyst's preferences, the nature of routine work, the extent of usage of the item, and of course, availability of funds. The following list of items of equipment is sufficient for most routine work. There will be additional tools other than those listed that you undoubtedly will feel are justifiable as your work progresses.

- Binocular microscope: Most routine work will be done at 7× or 15×. Contrary to the opinion of many beginners, you will generally do better working under as low a magnification as possible consistent with the capability to discern identifying characteristics. Area, depth of field, and resolution will be much better at lower magnifications.
- Illumination: A good adjustable light source is quite important. Halogen or tungsten lighting is preferable, although an inexpensive incandescent desk lamp is generally adequate.
- Blue "cobalt" glass: A 4 in. × 4 in. sheet as carrier and background contrast for the sample.
- White "milk" glass: A 4 in. × 4 in. sheet as carrier and contrasting background for the sample.
- Spot plates: Both black and white porcelain spot plates are useful to provide contrasting background for color development in spot testing. A "Harlequin" spot plate (black and white) is quite helpful in numerous

instances. A Pyrex glass spot plate may be desirable when heating is necessary.

- Forceps: These are indispensable to the analyst in manipulating the sample under the microscope. Because of the deteriorating effect of many fertilizer ingredients, stainless steel, nickel-plated, or Teflon tipped forceps are recommended. The latter are practically inert.
- Dissecting needles: Preferably stainless steel or nickel-plated.
- Small glass stirring rods: These are conveniently made by drawing small diameter solid glass rods to a small tip over a flame. These fine tipped rods are useful not only for stirring but to manipulate particles in a liquid spot test under the microscope.
- Semi-micro test tubes and support: 10 × 75 mm and 13 × 100 mm Pyrex tubes with support rack are convenient.
- Spatulas: Small porcelain and stainless steel or aluminum tools blunt and narrow tipped, 7–20 mm wide, are useful for transferring and manipulating samples.
- Mortar and pestle: Small 30–50 mm ID, porcelain is usually adequate. Hard materials may require one of mullite, alumina, diamonite, or agate.
- Evaporating dish: One or two Pyrex or porcelain, 50–70 mm diameter should suffice.
- Watchglasses: Several, of 40–100 mm diameter, where a concave surface is necessary to hold a sample for examination.
- Alcohol burner: Useful for heating some spot tests, speeding reactions, evaporations, sublimations, etc. A wick-type burner is inexpensive and generally satisfactory. Of course if a gas Bunsen burner is available, this might be used.
- Beakers and funnels: Several Griffin or Berzelius beakers, 50 to 250 mL. A glass or polyethylene funnel with a short 8-mm stem and one with a 14-mm diameter stem are useful.

As the microscopist advances and goes into tangent fields such as photomicrography, crystallography, etc., additional equipment becomes necessary. A photomicrographic set-up, ready for use when the necessity arises, can be extremely useful for recording samples for future reference. A petrographic microscope is warranted in some instances where considerable crystallographic work is done.

- Standard samples: Keep all the commonly encountered ingredients convenient. Specimen vials with tight screw caps are ideal containers. Deterioration is often a factor to consider with standards and photomicrographs are referenced. Additional samples will continually be added to your standards, so space and provision for expansion of samples should be available. To facilitate rapid selection of a standard sample, it is best to keep a reference list identifying samples by number. The vial caps are numbered and the vials themselves labeled by contents. The desired material can then be selected easily by number. When additional materials are added to the collection, the numbered vial caps can be changed among the vials and the list revised, thereby eliminating the necessity of changing the material in the containers.
- Reagents: Reagents for spot tests are conveniently kept in 1-oz dropper bottles (liquids) or wide mouth bottles (solids). Deterioration and reactivity of reagents must be considered and amber or polyethylene containers used where appropriate.

Identification of Organic Materials

The most important factor in the identification of organic materials in fertilizers is experience. The microscopist can read, listen to descriptions, and look at pictures of a particular material day after day; but none of this can substitute for the experience of actually seeing them through a microscope. Many organic particles have characteristics that may remind you of some common object; but as long as the characteristic peculiar to that material brings that association to mind, it is an aid to identification. Of course, the individual characteristic structural features must be learned and observed in identifications.

The sense of smell can play an important part in what the microscopist decides he or she is seeking. Even very small percentages of Peruvian guano, chicken manure, tung meal, or tobacco stems in a fertilizer can be detected through their characteristic odors. Of course, odors are often masked, and it is not the sole means of identification, but may serve as an indication leading to visual or chemical confirmation.

Often, it is desirable to roughly separate the organic from the inorganic ingredients of a mixture. This may be conveniently done by mixing the sample in a solvent such as carbon tetrachloride or chloroform (wear protective gloves and work under a fume hood) and allowing the organics to float to the top and the inorganics to sink. A rough percentage by weight may be estimated by using a weighed amount of sample, separating and drying, then weighing the fractions. If a few particles are present in the wrong fraction, these may be removed with forceps under the microscope.

Common Organic Fertilizer Ingredients

The following briefly discusses some of the commonly encountered organic fertilizer ingredients purposely omitted in some commonly found feeds. Coverage of these is noted in previous chapters. Many feed ingredients would make excellent fertilizer materials but are of too much value in the form of fertilizers, or in times of surplus, which brings down the price of otherwise too-valuable materials. Certain intensively grown crops such as shade tobacco, where quality commands a premium price at market, justify the use of fertilizer ingredients such as cottonseed meal, which would be economically unfeasible in the average fertilizer. Of course, availability of materials varies in different sections of the country, and this is often a determining factor. Many materials, because of inherent toxicities or other reasons of unsuitability, are not used at all for feeds and find a ready market in fertilizers.

Sewage Sludge. Known by various terms as activated sludge, activated sewage products, sludge tankage, sludge, etc., or trade names as Milorganite, among others. This material generally runs about 5% nitrogen, mostly in insoluble forms, and usually about the same percentage, or slightly less, of available phosphoric acid. The AAPFCO definition is as follows:

> "Activated sewage products are those made from sewage freed from grit and coarse solids and aerated after being inoculated with microorganisms. The resulting flocculated organic matter is withdrawn from the tanks, filtered with or without the aid of coagulants, dried, ground, and screened."

Under the microscope sewage sludge is quite easily distinguished from animal tankages. It appears as small, dull gray to black, irregularly spherical bits, sometimes aggregated often with some fine grit, with occasional to numerous obvious white fibers protruding from the particles. It is soft and easily crushed with forceps.

Tankage. This is a general term pertaining to any tankage of animal origin. AAPFCO defines the term:

> "Tankage (without qualification) is the rendered, dried, and ground byproduct, largely meat and bone, from animals (slaughtered or that have died otherwise)."

The raw sources of this material are quite varied. So-called digester tankage and process tankages, which consist of treated crude insert nitrogenous materials such as leather scrap, feathers, hair, wool waste, rags, garments, etc., have come to be generally included in this grouping. Under heat, pressure, and sometimes acid-induced hydrolysis, a conversion of keratin to purines and amino acids (or even to ammonia) is effected in the crude materials, resulting in a potentially usable form of nitrogen. Nitrogen in these materials may vary from 2 to 12% and the phosphoric acid from 1 to 15%, depending of course upon the raw material used. Some of the general types of tankage according to derivation are as follows:

Packinghouse Tankage. A by-product of a meatpacking house (mainly meat, blood, etc.). Sometimes called slaughterhouse or "high grade" tankage, this is recognizable as rather glossy usually brittle, dark or black particles, somewhat larger than sludge particles, which will generally break up with a crystalline appearance. It is not uncommon for bits of bone, dried blood, or hair to be present. The flesh particles look like just what they are—over cooked or charred bits of meat.

Leathermeal Tankage. This is a "process tankage" which is derived from leather scraps. Leathermeal tankage is treated separately here because of the potential detrimental horticultural or agronomic effect of chromium salts which may be present in that tankage, which is derived from chrome tanned leather scraps. This product has a characteristic leather odor which can usually be easily identified. The particles are relatively medium to large in size, black and somewhat glossy, brittle and, if chrome salts are present, these will be obvious as occasional small bright red particles.

It may be desirable to measure the chromium quantitatively which may be done by a spectrophotometric determination of the soluble red-violet chromate complex of diphenylcarbazide in acid solution. This method is quite sensitive and specific for chromium and molybdenum, the latter not likely being found in leather meal, but possibly in mixed fertilizers. See Sandell, 1950, p. 260.

Feathermeal Tankage. A hydrolyzed process tankage derived from poultry feathers and skin remnants. The feathers as such are not generally recognizable except for some of the feather shafts. These are medium to large size, buff to amber resinous appearance, nearly transparent, somewhat flattened cylindrical shape with a conspicuous longitudinal grove, and often with barbs on the sides. Occasional opaque flexible bits of skin may be found.

Other Process Tankages. These are quite varied in appearance. It may be a heterogeneous mixture of very fine, light particles, which frequently adhere to and obscure other (particularly inorganic hygroscopic) particles. Often it will be found composed of black or dark green cartilaginous or cutaneous material or rubbery or mealy lumps. Here again only experience and comparison with known materials will enable the microscopist to make identifications.

Bone Meal. Both raw and steamed bone meals are commonly used in fertilizers. The former is ground after being cooked in open vats to remove meat and fat. The latter is ground after being steamed under pressure. They generally contain around 2 to 3% nitrogen, 15 to 30% phosphoric acid, and 23 to 45% calcium oxide. It appears under the microscope as rock like, irregularly oblong shaped, dull flat white particles. They break up fairly easily under pressure of the forceps into slivers. Unlike most organic materials, bone meal will sink in carbon tetrachloride. An individual particle placed on a black spot plate with a drop of 1:1 hydrochloric acid will give off a slow steady stream of bubbles, but it is not nearly as vigorous a reaction as is the effervescence of carbonates in HCl. A drop of molybdate solution with a bone particle on a white spot plate will produce minute yellow crystals confirming a phosphate. To distinguish from defluorinated or rock phosphate particles, which they may occasionally resemble, add a drop or two of Millon reagent (Warm 5 g mercury in 25 mL conc. HNO_3 and add 50 mL H_2O.) Use mercury with extreme caution!! Decant supernatant for use after standing overnight (Store in dark bottle.) to particle on a white spot plate. Within 5 min or so, a pink to red color will develop from residual proteins in the bone, while the phosphate particles will disintegrate with no color change.

Tobacco Stems. Consists of ground waste tobacco products. These generally contain about 2% nitrogen, 1% phosphoric acid, and 6% to as much as 13% potash. Bits of ground stem are elongated, fibrous, with "broken" ends, somewhat resembling the vegetable fibers found in cattle and sheep manures, but are generally of a lighter buff color. Flake-like leaf particles are often found with the stem bits. The odor of tobacco stems in a mixed fertilizer is usually recognizable, though sometimes masked by more the potent odors present.

Cotton Burr Ash (Cottonseed Hull Ash or Carbonate of Potash). Ash from well burned cottonseed hulls, used particularly in tobacco fertilizers. Analyses of this product vary widely, with average values being 5% phosphoric acid, 25% potash, 10% calcium oxide, and small amounts of various micronutrients. Under the microscope it appears as non-crystalline, somewhat conglomerate, variously shaped particles, off-white to blue-gray heterogeneous in color, similar in appearance to superphosphate but grayer and not as rounded. Minute particles are usually present, but larger particles should be picked out for examination and spot testing. A particle in 1:1 HCl on a black spot plate will effervesce violently, with a definite sulfide odor given off distinguishing it from limestone materials, which also effervesce. It may be checked for potassium as follows: after effervescing ceases, add a few drops 20% sodium hydroxide, 1 drop 37% formaldehyde, and 1 drop 1% sodium tetraphenylboron, stirring between each addition. A white precipitate confirms potassium.

Cottonseed Meal. By-product of cottonseed oil manufacture composed principally of cottonseed with some hull, analyzing 6 to 7% nitrogen. It is esteemed as a source of insoluble nitrogen, particularly in shade tobacco fertilizers. High-grade meal is a deep yellow color. If fermented it is more of a red-brown, and if moldy, gray-brown. If a high percentage of hulls are present the color will be dark brown. Under low magnification, pick out a few hulls. They will be light to very dark brown, curled (concave inside), tough and leathery. In cross-section, longitudinal layers (amber and yellow or amber-yellow-amber) are apparent with lateral edge striations. Cotton fibers or linters are often found adhering to the outer surface of the hull. Under oblique illumination, these linters appear silvery transparent and the amber contrast of the hull particle is better resolved. The meal itself, generally a bright yellow-brown, may contain oily chunks of kernel with numerous reddish pigment cells.

Peanut Hull Meal. Generally used as a conditioner rather than a source of plant foods, of which it usually contains fractional percentages. Under low magnification the buff colored hull particles are easily identified by the "mountainous" ridged appearance of the outer surface. Layers of ridged surface can be seen at right angles to each other. Bits of the red skin are often seen. Stem fragments may be distinguished by the internal dark longitudinal resinous deposits. As a rule the characteristic hull particles are all that are necessary to make an identification.

Rice Hulls. As with peanut meal, rice hulls are generally used as a conditioner, being light and bulky. The thin, elongated, broken, bright yellow hulls with characteristic uniform shiny, ridged striations and precise cross-hatching make this product quite easy to identify.

Castor Pomace. Ground residue of castor beans after oil extraction, containing about 5% nitrogen, 2% phosphoric acid, and 1% potash. Hulls are dark brown to amber, except the slightly convex outer surfaces, which often has mottled light and dark surfaces. Inner surfaces are dark amber and appear dimpled. Castor pomace is thicker than soybean, but thinner than tung meal, with definite edge striations. Sharply angled pieces resemble broken formica. The meal particles, contrasting with the dark hulls, are buff to off-white in color, sometimes resembling bone meal; but easily distinguished from the latter by the manner in which they break up with the forceps, crushing rather than splintering. In addition, castor particles will float in carbon tetrachloride, while bone will sink.

Tung Pomace. Ground residue of tung nut after oil extraction. Typical analysis is about 4% nitrogen, 2% phosphoric acid, and 1% potash. Hulls are dark amber or brown. Definite edge striations, but much thicker than castor pomace and with a ridged, "mountainous", rather than smooth outer surface. Generally broken into "chunks" in which the thickness may be as great as the other two dimensions. Occasional samples are encountered of tung pomace which has been solvent extracted. In this the structural features are the same of course, but the color is much lighter, pale buff to cream color. The meal is yellow resembling cottonseed meal, except that no fibers will be found as in the latter. As with most meals, identification is dependent upon recognizing the hull rather than the meal particles.

Soybean Meal. The ground residue of soybeans from which the oil has been extracted. It is not too commonly found in fertilizers, generally being of more value in feeds. Average analysis runs about 6% nitrogen, 1% phosphoric acid, and 2% potash. The hull is quite characteristically thin in cross section and curled, even rolled up, yellow in color. The outer surface is glossy with minute depressions, the inner surface white and spongy.

Guano. Often listed as Peruvian Guano or bird guano, it is the partially decomposed excrement of birds (or of bats or other animals). Generally contains about 12% nitrogen, 11% phosphoric acid, and 2% potash, although this varies widely. Easily identified visually by its soft texture, buff to tan color, and characteristic feather particles. These feather particles are characteristic in themselves, differing markedly from raw or processed poultry feathers. Often, in a mixture with only a small amount of guano present (which incidentally, occurs frequently), the bits of excrement are minute and far between, but the feather particles are a dead give-away. The odor, once the analyst learns to recognize it, is characteristic.

Manures. Cattle and sheep manures may be difficult to distinguish from each other, but with a little experience, they may be differentiated by the characteristic bovine or sheep hairs generally present. Abundant semi- or undigested vegetable fiber will be found along with the dark amorphous manure itself. Odor is characteristic, particularly with poultry manure. The latter is easily distinguished by the presence of grit and identifiable particles from poultry feed.

Peat and Muck. Sometimes used as filler or conditioner, peat and muck has little value as a source of plant food, containing an average of 1.5% insoluble nitrogen. It will hold about six times its own weight in moisture, hence it is of value in chemical fertilizer mixtures. Peat is partly decomposed plant residue and portions of recognizable plant structure can be found. Muck is the same thing in an advanced stage of decomposition and may be nearly amorphous. They will be medium to very dark brown, soft, and easily crushed.

Urea [$CO(NH_2)_2$]. The commercial synthetic product runs about 42% (all water-soluble) nitrogen. Used extensively in fertilizers as well as feeds, urea is found naturally in Peruvian guano and poultry manure. If present in pelleted form, it is easily identified by its ice-like appearance and hygroscopicity. Often a loosely formed lump of organic materials will be found with a small bit of hygroscopic urea as the nucleus. If a particle is suspected to be urea, dropping it into carbon tetrachloride will, if it floats, differentiate it (and urea form) from inorganics. Check a particle on a white spot plate to confirm the absence of ammonium ion with Nessler's reagent. If negative, dissolve a particle in water on a white spot plate. (If nearly insoluble, it may be ureaform). If positive, dissolve a particle in water, add a drop of dilute NaOH, heat slowly to dryness, re-dissolve residue in a drop of water, then add a minute amount of urease. After 2–3 min, check again for ammonia with Nessler's reagent. If positive, confirms urea (or a less intense indication may mean ureaform. Ureaform will char in a test tube over heat, while urea will not). Another good spot test for urea (and again, less intensely, ureaform) is to place a particle on a white spot plate and add 1–2 drops of Dimethylaminobenzaldehyde solution. After a minute or two, a bright yellow condensation product is a positive indication (as long as copper is absent). Urea and ureaform may not be identifiable visually, since they are sometimes present on a "carrier", such as organic meal, vermiculite, etc. Urea may be present in an "ammoniated" superphosphate, in which case it may be more difficult to identify.

Urea-formaldehyde (Ureaform). Approximately 38% mostly insoluble nitrogen, ureaform is made by reacting urea with formaldehyde to produce a nearly insoluble but slowly available source of nitrogen. Grades used in fertilizers must have a minimum nitrogen activity index. If not in a "carrier" it may be seen as white, gray, or dyed colored lumps. Often appears as aggregated cubic "crystals," somewhat like muriate of potash but with more of a waxy plastic appearance than crystalline. They will break up easily under forceps, being relatively soft. See additional discussion in preceding paragraph with urea.

The preceding list of organic materials is limited both in scope and intensive description. A discussion of all the organic ingredients encountered in the various parts of the country would fill a textbook. Many other ingredients are covered in publications such as those listed as suggested references in the back of the manual. Concerning recognition and identifications, there is no substitute for experience, the individual microscopist making their own analogies as they recognize and learn characteristics of the various materials.

Identification of Inorganic Materials

The first steps in identifying inorganic materials are to make a tentative visual identification followed by confirmation of the cation and anion present through a series of chemical spot tests.

Materials

The following is a list and brief description of some of the more commonly encountered sources of the three primary and several secondary plant foods. However, some forms of materials may vary in color, particle size, chemical analysis, and appearance, depending upon the method of manufacture.

Ammonium Sulfate. Colorless or grayish, transparent, hexagonal, or rhombic crystals, 0.2–4.0 mm, fairly brittle. Generally quite characteristic, long, hexagonal in cross section, pencil shaped appearance. 20 to 21% nitrogen in ammoniacal form.

Sodium Nitrate (nitrate of soda, chile saltpeter). Usually pelleted, 1–3 mm, white opaque waxy, sometimes hollow spheres. Crush readily into minute, deliquescent crystals. 16% nitrogen in nitrate form.

Calcium Nitrate. White, opaque pellets or chunks, 2–8 mm, granular glossy surface, very easily crushed into minute, deliquescent crystals. About 15% nitrate nitrogen and 27% CaO. Often contains up to 1% ammoniacal nitrogen as a consequence of the process of manufacture.

Ammonium Nitrate. White, opaque, imperfectly shaped spherical, sometimes hollow pellets or chunks, 2–5 mm with a dry powdery surface. Easily broken into small hygroscopic crystals. 33% nitrogen, half of it ammoniacal, half nitrate form.

Ammonium Sulfate-nitrate. Off-white to buff or yellowish, semitransparent, brittle 1–3 mm aggregated crystals. The color and structure are quite characteristic and generally readily recognizable. 26% nitrogen, one-fourth of it nitrate and three-fourths ammoniacal.

Ammonium Nitrate Limestone. Buff to tan or off-white, fairly smooth surfaced, 2–5 mm irregular lumps, difficult to crush into opaque white powder. Sometimes erroneously termed calcium ammonium nitrate, it is actually a mixture of ammonium nitrate and finely pulverized limestone or dolomite. About 20% nitrogen, half ammoniacal and half nitrate forms.

Diammonium Phosphate. Crystals are 1–2 mm, slightly grayish, transparent, and platelike. Pellets (more common form) are irregular conglomerate, 1–4 mm, off-white, grayish, or buff. Similar in appearance to superphosphates, but harder and more of a crystalline appearance. About 20% ammoniacal nitrogen and 50% P_2O_5.

Superphosphate. Off-white to gray, buff or tan, small non-crystalline granules from a few hundredths of a mm up in size. More often found as conglomerate lumps, 1–6 mm, easily crushed into smaller particles. About 20% available P_2O_5.

Double or Triple Superphosphate. Essentially the same appearance as ordinary superphosphate, distinguishable by comparative spot testing for gypsum ($CaSO_4$) which is prevalent in ordinary superphosphate but nearly absent in triple superphosphate. About 46% available P_2O_5.

Ammoniated Superphosphate. Generally same appearance as other superphosphates, but may have a more glazed over surface appearance or a slight blue-green tint. Made by impregnation of superphosphate with various ammoniating solutions such as ammonium nitrate, urea, ammonia liquor, anhydrous ammonia, etc., resulting in a product of varying percentages and forms of nitrogen.

Magnesium Ammonium Phosphate. Dull, bone-white, opaque, easily crushed irregular, generally rounded granules, 2–5 mm. 8% ammoniacal nitrogen, 40% P_2O_5, 24% MgO.

Ammonium Polyphosphate. Dull, white, semitranslucent with opaque white coating, hard, very irregular, rounded aggregates, 2–4 mm. 15% ammoniacal nitrogen, 60% P_2O_5.

Potassium Polyphosphate. Dull, white, semitranslucent, hard rock-like, sharply angled chunks of aggregated crystals, 2–7 mm. 46% P_2O_5, 38% K_2O.

Potassium Nitrate. Crystals are rhombic, colorless, transparent, 0.1–0.5 mm. Generally pelleted, pinkish to white spheres, 1–3 mm, may be hollow, somewhat difficult to crush. Similar in appearance to sodium and ammonium nitrates, but less hygroscopic, distinguishable through spot tests. About 13% nitrate nitrogen and 44% K_2O.

Potassium Chloride (muriate of potash). Very characteristic cubic crystals 0.1–5.0 mm, colorless or pink-orange splotched. When crushed under gentle pressure, generally cleave along cubic crystalline lines. Approximately 60% K_2O, 46% chlorine.

Potassium Sulfate (sulfate of potash). Rhombic or hexagonal, white or pink tinged, semitransparent to opaque crystals, 0.1–3.0 mm, often in irregular aggregates which crush easily into the smaller crystals. Approximately 50% K_2O.

Sulfate of Potash-magnesia (Sul-Po-Mag). Medium to large, 0.5–6.0 mm, white to pink-orange tinged, frosty, translucent, irregular crystals. Somewhat similar in appearance to potassium chloride, but lacking the latter's characteristic cubic shape and considerably harder to break with the forceps. The domestic form contains approximately 21% K_2O, 18% MgO.

Magnesium Sulfate (emjeo, epsom salts). Frosty white to transparent or clear, irregular monoclinic crystals, 0.1–2.0 mm. Emjeo, about 30% MgO, appears similar to Sul-Po-Mag, but crystals are smaller with no pink tinge, and may be differentiated by spot testing for potassium. Epsom salts, about 18% MgO, is a more hydrated form, with clearer, longer crystals. Both are soluble forms.

Manganese Sulfate. Fine, opaque white particles, 0.1 mm or less. It is soluble and averages 32% MnO.

Manganese Oxide. Very fine, dark brown powder. Insoluble forms, averages 60–65% MnO. Some products may run as high as 70–80% MnO.

Copper Sulfate. Medium to large, 0.3–3.0 mm, soluble, triclinic crystals, conspicuous by their characteristic beautiful blue color. Approximately 31% CuO.

Copper Oxide. Very fine, opaque, dull black powder, darker than manganese oxide. Insoluble, domestic fertilizer grade will probably average 80% CuO.

Ferrous Sulfate (iron sulfate). White opaque to pale blue-green clear, monoclinic, efflorescent, characteristically rhombohedral crystals, 1–3 mm. Generally acquires a yellow-brown to reddish-brown oxidized coating of basic ferric sulfate in moist air or mixed fertilizers. Approximately 28% Fe_2O_3.

Chelated Iron. Generally not present in crystalline form, but impregnated in vermiculite particles, which are usually reddish-brown. (See vermiculite description below). Typically, 6-12% Fe_2O_3.

Borax (sodium borate). Clear, vaguely green tinted, hexagonal efflorescent plate-like crystals, 0.3–1.5 mm, 35–45% B_2O_3.

Zinc Sulfate. Fine, opaque, white powder, generally less than 0.1 mm. Ranges from 27% to 45% ZnO, depending upon the state of hydration. (Three hydrated forms are commonly used).

Aluminum Sulfate. White or slightly yellowish, irregular translucent, crystals, 1–5 mm. Frequently found in acid-forming fertilizers such as azalea-camellia or blueberry formulations. Approximately 15% $A_{12}O_3$.

Cobalt Sulfate. Fine pink to red, opaque or translucent crystals, usually 0.1–1.0 mm, sometimes aggregated. Approximately 25% CoO.

Sodium Molybdate. Fine, colorless, clear, plate or sheet-like crystals up to 0.5 mm. About 56% MoO_3.

Sulfur. Bright yellow, usually large brittle crystals, up to 3 or 4 mm. Occasionally found as black or black-coated particles in lower grade material. Easily recognized by burning a particle and noting the blue flame and evolution of the pungent sulfur dioxide.

Limestone (calcium carbonate). Fine white or off-white powder, occasionally present as granules up to 5 mm. May be quite prevalent when used as a filler in mixed fertilizers. Effervesces strongly in dilute HCl, with a positive calcium and negative magnesium spot test. Approximately 90% $CaCO_3$.

Dolomite (dolomitic limestone). Generally same appearance as limestone, except buff to tan color, exhibits weaker effervescence, and shows positive calcium and magnesium spot tests. Generally about 50% $CaCO_3$ and 35% $MgCO_3$.

Basic Slag. Very finely divided, dark brown with occasional lighter and darker particles. Analyses will vary greatly depending upon origin, averaging about 15% available P_2O_5, 45% CaO, 20% Fe_2O_3, smaller amounts of magnesium, manganese, and other secondary elements.

Phosphate Rock. The ground form is similar in appearance to superphosphate, but more powdery. Sometimes found as stone-like lumps or chunks, having been crushed, but not finely ground. May be used as a filler or possibly present in poor grade superphosphates. Reacts with HCl and turns yellow with molybdate solution, (as do superphosphates), but not with silver nitrate. Contains 25–35% P_2O_5 (mostly insoluble) and 40–50% CaO.

Pebble Phosphate, Phosphate Spiral or Flotation Concentrates, etc. Tan and white opaque pebbles, 0.2 up to 10.0 mm, with a small amount of sand. Fossilized bone and tooth structure often apparent.

Phosphate Tailings. Heterogeneous, with abundant sand, often black, worn, pebble-like particles, remnants of fossilized bone and tooth structure, varying size.

Colloidal Phosphate. Very finely divided, buff, yellowish tan color, generally clings in soft lumps. Usually high (9–12%) in iron and aluminum oxides. About 2–3% available P_2O_5, and 16% or more insoluble P_2O_5.

Vermiculite. Heated expanded magnesium mica, extremely light weight, and absorptive. Easily recognized by its "accordion bellows" appearance, silvery tan color, from 0.5–3.0 mm. Used frequently as a conditioner, or a carrier for chelated iron, urea, ureaform, etc., which may alter the color.

Combinations or various fertilizer ingredients may be found in pellet or granular form. One common example of this is ammoniated superphosphate. Innumerable combinations may be encountered, varying in color, shape, and form. In many of these cases the usual crystalline form will not be present to aid in identification, and the microscopist will have to depend upon familiarity with known materials and spot tests.

Spot Testing of Inorganic Compounds

Some simple spot tests used in identifying the more frequently encountered inorganic compounds in mixed fertilizers are covered. Many other spot tests can be used in addition to or in place of these and may be found in publications such as "Feigl's Spot Tests in Inorganic Analysis." Those which follow are those which are most useful or potentially so in routine identifications. More than one test may be desirable in confirming difficult identifications. These tests may be and in some cases must be varied to meet individual circumstances.

Select a clean, typical crystal, pellet or particle and break in into smaller pieces with forceps or mortar and pestle to expose surfaces free of any possible contamination. It is often desirable to clean the particle by rolling it between the thumb and forefinger or in damp or dry tissue paper, or by rinsing with distilled water or carbon tetrachloride. (Wear protective gloves and use in a fume hood.)

Reagents

Regarding the reagents used in spot testing, those with no percentage concentration indicated are the common concentrated commercial product. Those with concentrations given are approximate percentages and in aqueous solution unless otherwise specified (always use distilled or deionized water). Reagent solutions which must be made up otherwise are as follows (Wear protective gloves and use in a fume hood.):

Molybdate Solution. Combine 15 mL NH_4OH and 27 mL H_2O, and 10 g MoO_3, and allow to cool. Add 50 mL HNO_3 to 115 mL H_2O and allow to cool. Add the first solution slowly, while stirring, to the second solution. Allow to settle several days and decant the clear supernatant solution. To 100 mL of this, add 10 mL of a 10% NH_4NO_3 solution. Store in a dark bottle.

Nessler's Reagent. Dissolve 14 g NaOH in 90 mL H_20 and filter if necessary. Add 5 g red HgI_2 and dilute with H_2O to 100 mL. Mix well, allow to settle and decant the clear supernatant for use.

Uranyl Zinc Acetate. Dissolve 20 g uranyl zinc acetate in 25 mL warm 15% acetic acid and dilute to 100 mL or it may be made as in "Feigl's Spot Tests in Inorganic Analysis" (Feigl, 1958, p. 230).

Benzidine. Dissolve 5 g benzidine in 100 mL 10% acetic acid.

Phenylhydrazine. Mix one part phenylhydrazine with two parts glacial acetic acid.

DMAB. Dissolve 2 g dimethylaminobenzaldehyde, $(CH_3)_2NC_6H_4CHO$, in 90 mL methanol (CH_3OH) and 10 mL HCl. Store in amber bottle.

Anions

Chlorides. Add 1 drop 10% $AgNO_3$ to particle on black spot plate. An expanding chalk-white precipitate forms immediately.

Sulfates. Add 1 drop 10% $AgNO_3$ to particle on black spot plate. Small white needle crystals form on the particles and will dissolve upon addition of 1 drop 1:1 HNO_3. Dissolve a particle on black spot plate in a drop or two of water plus 1 drop 1:1 HCl. Add 1 or 2 drops 10% $BaC1_2$. A fine white precipitate appears.

Phosphates (mono- and di- basic phosphates as found in superphosphates). Add 1 drop 10% $AgNO_3$ to particle on white spot plate. Bright yellow needle crystals form on the particle. If a tri-basic phosphate (rock phosphate or bone) is suspected and shows negative to $AgNO_3$, add a drop of molybdate solution. Minute yellow crystals form.

Nitrates. Add 1 drop 10% $AgNO_3$ to particle on black spot plate. The particle will dissolve without further apparent reaction. To a particle on a white spot plate, add a minute amount of Brucine alkaloid powder, then 1 drop concentrated H_2SO_4. A brilliant red color appears.

Carbonates (as dolomite, limestone, or cotton burr ash). Add 1 drop 1:1 HCl to particle on black spot plate. A vigorous effervescence occurs.

Cations

Ammonium. Add 1 drop Nessler's reagent to particle on white spot plate. A chalky red-brown formation surrounds the particle.

NOTE: In a mixed fertilizer, many particles may become contaminated from ammonium salts present, and this must be considered when a positive Nessler's test is encountered.

Calcium. Dissolve in a drop or two of 1:1 HCl on black spot plate. Add 1 drop saturated solution ammonium oxalate, then 1 or 2 drops NH_4OH. An intense white precipitate forms.

Sodium. Dissolve a particle in 1 drop H_2O, add 1 drop acetic acid, 1 drop ethanol, then 1 or 2 drops uranyl zinc acetate. A green-yellow precipitate slowly forms. It may be observed to advantage under UV light, if difficulty is encountered in discerning the precipitate.

Potassium. Dissolve a particle in H_2O on black spot plate. Add 1 drop 20% NaOH, 1 drop 37% HCHO, and 1 drop 1% sodium tetraphenylboron. A white precipitate appears. Dissolve a particle in H_2O on black spot plate, add 1 drop 10% $AgNO_3$, 1 drop 5% sodium cobaltinitrite solution. A yellow precipitate forms.

Magnesium. Dissolve particle in 1 drop H_2O on white spot plate. Add 1 drop 20% NaOH, 1 drop 0.05% alcoholic quinalizarin solution; compare blue color with violet color of blank after standing awhile. Dissolve particle in 1 drop H_2O on white spot plate. Add 2 drops 20% NaOH, 1 drop 1% clayton yellow solution. A red color or precipitate appears.

Manganese. Dissolve particle in H_2O on white spot plate. Add 1 drop 20% NaOH, 1 or 2 drops benzidine. A blue precipitate appears. Dissolve particle in 2 or 3 drops H_2O in a semi-micro test tube. Add 1 drop concentrated H_2SO_4, 1 drop 10% $AgNO_3$, and a few tiny crystals KIO_4. Heat slowly over alcohol burner. The purple permanganate color develops.

Zinc. Dissolve a crystal in 1 mL H_2O in a semi-micro test tube. Add 1 drop acetic acid, 2 drops 30% $Na_2S_2O_3$, and 2 to 3 mL 0.1% diphenylthiocarbazone in $CC1_4$ solution. Cork the tube and shake vigorously. The green diphenylthiocarbazone in the $CC1_4$ phase will turn red-purple. This is an extremely sensitive test, so precautions against contamination should be observed.

Aluminum. Dissolve particle in a drop of H_2O on white spot plate, add 1 drop NH_4OH, 1 drop 0.05% alcoholic quinalizarin solution, then acidify with acetic acid. A red lake will form.

Copper. Dissolve a particle in a drop of 1:1 HCl on white spot plate. Add several drops tetraethylenepentamine and stir. An intense blue complex will form. Extracting a copper solution into diphenylthiocarbazone in $CC1_4$ as described above in the zinc test will produce a yellow-brown complex.

Molybdenum. Dissolve a particle in a drop of H_2O on white spot plate. Add 1 drop 10% KSCN and 1 drop 1:1 HCl. If ferric ion is present, red color will appear. Add 1 drop 30% $Na_2S_2O_3$ to destroy this and a brick-red thiocyanate complex of molybdenum appears. Dissolve particle in 1 drop H_2O on white spot plate. Add 1 drop acetic acid, 1 drop phenylhydrazine solution. A blood red color or precipitate appears.

Cobalt. Dissolve particle in H_2O on white spot plate. Add 1 drop ethanol and 1 or 2 drops 30% $Na_2S_2O_3$ to produce a blue color. Dissolve particle in 1 drop 1:1 HCl on white spot plate. Add 1 or 2 drops of a 10% NH_4SCN in acetone solution to produce a deep blue or blue-green color.

Iron. Dissolve a particle in a drop of 1:1 HCl on white spot plate. Add 1 drop 5% $K_4Fe(CN)_6$. An intense deep Prussian blue

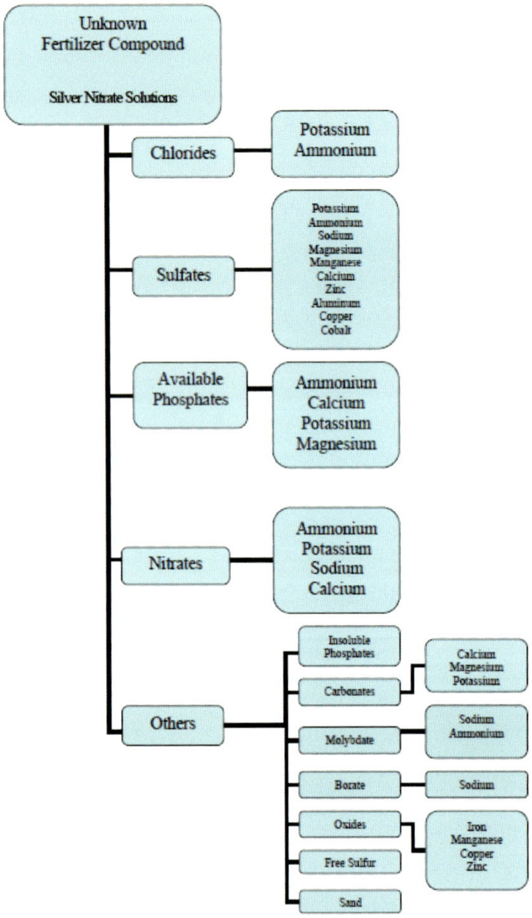

Fig. 6.1. A suggested spot-testing procedure in identification of common inorganic materials found In fertilizer mixtures.

color confirms ferric ion while a lighter, milky blue-green Turnbull's blue indicates ferrous ion, the former being a more highly oxidized form of the latter. Chelated iron as commonly found on vermiculite particles will give the Prussian blue color. Dissolve a particle in a drop of 1:1 HCl and a few drops of water. Add a drop of 10% KSCN. A blood red color confirms ferric ion, which may be destroyed upon addition of a drop or two of 30% $Na_2S_2O_3$.

Boron. To a crystal on a white spot plate, add 1 drop saturated solution of $HgC1_2$ to produce a red precipitate. This reaction will occur only in a neutral environment, so precautions against possible contamination must be observed.

Qualitative Microscopic Analysis of a Mixed Fertilizer

The following procedure is an example of a routine that a control laboratory would use in identifying ingredients of a mixed fertilizer. From the manufacturer's tag accompanying the sample, we find guaranteed percentages of total, ammoniacal, nitrate, water-soluble, and water-insoluble nitrogen; available phosphoric acid as P_2O_5; water-soluble potash as K_2O; copper as CuO, and iron as Fe_2O_3.

Spot Test

Quantitative chemical analyses will reveal whether or not it contains amounts within tolerances conforming to law. We are concerned here only with qualitative identification of the ingredients. The tag states that these guaranteed percentages are derived from ammonium nitrate, ammonium sulfate, superphosphate, muriate of potash, urea, guano, activated sludge, castor pomace, tung meal, copper sulfate, and iron sulfate. We can now follow a systematic identification of these materials visually and by means of spot tests as follows:

Activated Sludge. These are very easily identified as black irregularly shaped, generally nearly spherical, fairly uniform bits, often containing fine bits of grit. Look for the protruding white fibers.

Peruvian Guano. In this particular sample, the guano is fairly prevalent and dry; therefore, it may coat and obscure other particles. Look for the feather particles protruding from buff colored irregularly shaped pieces. Crush some of the pieces with the forceps and notice the soft consistency and the frequent minute feather particles within the pieces of guano.

Castor Pomace. Look for the thin, slightly concave amber-brown, sharply angled pieces with obvious edge striations. They resemble pieces of broken formica. On the outer surface of some, you will notice a mottling of light and dark brown color. The thinness and this mottling along with the general shape will distinguish castor from tung pomace.

Tung Meal. These particles are easily distinguished as thick, dark amber chunks, broken along the lines of quite apparent edge striations. The outer surface (perpendicular to the striated edge) will be found ridged, rather than a smooth surface.

Ammonium Nitrate. This is generally found in the pellet form as it is here. Look for the fairly large white, opaque spheres, sometimes slightly flattened and/or with a hole in the side or partially hollow. The urea in this mixture is also present in pellet form, but it can be quite readily distinguished from the ammonium nitrate by its ice-like appearance. The urea also is usually more densely coated with the dust from the guano and castor pomace present here. With the forceps, break apart a particle of both the ammonium nitrate and urea. Notice the crystalline structural differences, the relative opacities. The reagents necessary for spot testing are:

1. Silver nitrate, $AgNO_3$, 10%
2. Nessler's reagent
3. Brucine alkaloid, powder
4. Sulfuric acid, H_2SO_4, concd

Pick out and clean a pellet. Break it into small pieces with the forceps. Place some on a black spot plate and add a drop of $AgNO_3$. Note the solution of the particle. Place another particle on a white spot plate, add 1 drop Nessler's reagent, and note the reaction (positive ammoniacal test). Place a particle on a white spot plate, add a minute amount of brucine alkaloid, 1 drop of H_2SO_4, streak it slightly with a glass stirring rod, and note the intense red color development (positive nitrate).

Ammonium Sulfate. This is the only sulfate other than copper in this mixture and is easily picked out by the fairly large, hexagonal, oblong shape. Break one apart with forceps; note the breaking and transparency reagents:

1. Silver nitrate, $AgNO_3$, 10%
2. Nessler's reagent
3. Hydrochloric acid, HCl, 1:1
4. Barium chloride, $BaCl_2$, 10%

Place a broken piece on a black spot plate with $AgNO_3$ and note the minute, white needle-like crystals forming. Test for ammonium with Nessler's reagent as described above under ammonium nitrate. Dissolve a particle in 1:1 HCl on a black spot plate, add a drop of $BaCl_2$, and note the white precipitate (positive sulfate).

Superphosphate. Pick out a few of the large, buff to tan or gray, rough surfaced conglomerate particles. This appearance is typical of ordinary, triple, or ammoniated superphosphates, although the latter may have a smoother surface. Break a piece with forceps and note the internal construction. Reagents:

1. Silver nitrate, $AgNO_3$, 10%
2. Hydrochloric acid, HCl, 1:1
3. Ammonium oxalate, $(NH_4)_2C_2O_4$, saturated solution
4. Ammonium hydroxide, NH_4OH, concd

Place a broken particle on a white spot plate with a drop of $AgNO_3$ and note the gradual formation of bright yellow crystals (positive phosphate). Ordinary superphosphate contains gypsum as a product of the reaction in the manufacturing process. To distinguish it from, e.g., triple superphosphate, we can check for sulfate from the gypsum with $BaCl_2$ as described above under ammonium sulfate. Here, let us check for calcium to distinguish from diammonium phosphate, which has a similar appearance. Dissolve a piece as much as possible in 1 or 2 drops 1:1 HCl on a black spot plate. Add 1 drop $(NH_4)_2C_2O_4$ and 1 or 2 drops NH_4OH to make just alkaline. Note the white precipitate.

Potassium Chloride. Isolate a few of the large, cubic, colorless transparent or pink-orange splotched crystals. Break up a crystal and note the cleavage. Reagents:

1. Silver nitrate, $AgNO_3$, 10%
2. Distilled water
3. Sodium hydroxide, NaOH, 20%

4. Formaldehyde, HCHO, 37%
5. Sodium tetraphenylboron, (STPB), $NaB(C_6H_5)_4$, 1%

Place a broken crystal on black spot plate with a drop of $AgNO_3$. Note the intense chalk-white formation surrounding the crystal (positive chloride). Dissolve a piece in H_2O on black spot plate, add 1 drop NaOH, 1 drop HCHO, 1 drop STPB. The white precipitate confirms potassium.

Urea. The translucent ice-like pellets will likely be found covered with dust bits from the castor pomace and guano in this mixture. (See additional discussion under ammonium nitrate, above). Pick out a few particles and roll between the thumb and forefinger to clean them. Reagents:

1. Carbon tetrachloride, CCl_4
2. Dimethylaminobenzaldehyde, (DMAB)
3. Nessler's reagent

Break up a pellet to eliminate trapped air, as this will affect the specific gravity. Drop a piece into CCl_4. If it floats it is likely urea; if not you probably have ammonium nitrate. Compare the two visually again, and recheck the relative specific gravities. Place a pellet on a white spot plate, crush it, and add 1 drop DMAB. Observe for a minute or two and note the brilliant yellow color forming around the edge of the depression, eventually covering it. Copper may interfere with the DMAB reaction with urea. The copper in this mixture is in the sulfate form, which is readily discernible because of its color; therefore, it should present no problem to clean the urea pellet of any copper contamination. If any doubt exists however, the following procedure may be followed to confirm the urea. First, clean a pellet or two and dissolve in water. Add a drop or two of dilute NaOH and heat slowly to dryness. This will eliminate any ammonium ion contamination which would interfere. Dissolve the residue in a drop or two of water, stir in a small amount of urease powder, and allow to set at room temperature a few minutes. The addition of a drop of Nessler's reagent should turn a rust color, confirming the urea.

Copper Sulfate. This compound is probably the easiest of the minor elements to pick out in a mixed fertilizer. The fairly large, beautiful blue crystals can hardly be missed in the mixture. Even with a few hundredths of a percent present, it may be quickly found. Reagents:

1. Hydrochloric acid, HCl, 1:1
2. Tetraethylenepentamine, (TEPA), $NH_2(CH_2CH_2NH)_3CH_2CH_2NH_2$
3. Barium chloride, $BaCl_2$, 10%

Dissolve a piece from a broken crystal in 1 drop HCl on a white spot plate. Add 2–3 drops TEPA and stir. Note the intense blue color (positive copper). Dissolve another piece in 1 drop HCl on black spot plate and add 1 or 2 drops $BaCl_2$. The white precipitate confirms the sulfate ion.

Iron Sulfate. As a rule, in the form the material is placed in the mixture fertilizer (a hydrated ferrous form), it is a pale blue-green, translucent color. However, if the mixture has been standing for any length of time, the surface of the particles oxidizes to the ferric state in which case they will be opaque white to rusty colored, probably with a pale green to white interior. The generally rhombohedral shape along with the color renders it easily recognizable in most mixtures. Reagents:

1. Hydrochloric acid, HCl, 1:1
2. Potassium ferrocyanide, $K_4Fe(CN)_6$, 5%
3. Barium chloride, $BaCl_2$, 10%

Dissolve a particle in a drop of HCl on a white spot plate and add a drop of ferrocyanide solution. The intense Prussian blue color is positive confirmation of iron. Dissolve another particle in a drop of HCl on a black spot plate and add a drop or two of $BaCl_2$. The white precipitate confirms sulfate ion.

7 Weed Seeds of Agricultural Importance

Introduction

By definition, a seed is a plant at rest. A weed is a plant that is not valued where it is growing. And, a noxious weed is a plant that is difficult to control with normal cultural practices.

The identification of weed seeds is important to the agricultural microscopist for the following reasons: (1) detection of seeds containing natural toxicants (e.g., *Crotalaria*, Jimsonweed); (2) detection of seeds considered to be serious pests (e.g., Field Bindweed, Johnsongrass); (3) detection of seeds that may cause mechanical injury to animals that eat them (e.g., Cocklebur, Serrated Tussock); (4) detection of seeds that may impart objectionable odor or flavor to animal products (e.g., Wild Garlic, Field Pennycress); and (5) detection of seeds that may cause animals to limit feed intake or to refuse feed altogether (e.g., Field Pennycress, Corncockle).

To become familiar with all of the weed seeds occurring in feeds would take years of study. However, the microscopist should learn to identify those seeds containing toxic compounds and those that are considered noxious. Each state has a list of those seeds that have been declared noxious according to the state seed control laws. The microscopist should obtain the list(s) from those states that are pertinent.

The Association of American Feed Control Officials (AAFCO) has restricted the number of whole weed seeds that can occur in screenings used for feed. "Screenings must contain not more than four whole prohibited noxious weed seeds per pound and must contain not more than 100 whole restricted noxious weed seeds per pound." These weed seed species are specified by state seed laws.

A good reference collection is essential in identification and verification of weed seeds found in feeds. This collection may be acquired by selecting seeds from routine feed samples, by making field trips in the fall and winter, or by obtaining grain and agricultural seed screenings which usually abound in a number of different types of weed seeds.

The identification of weed seeds is based on the same characteristics that are used in the identification of feed ingredients. These include shape, size, color, texture, luster, and odor. The following photomicrographs and descriptions are offered as an aid to identifying weed seeds that may be encountered in routine analysis of feeds. They are arranged in alphabetical order by common name and include text and a map to tell its general distribution signified by darkened areas.

Weed Seeds

Fig. 7.1. Ballmustard (Mustard Family) [*Neslia paniculata* (L) P. Beauv]. Other common name: Yellow Weed. Pod oval to round in outline. Surface covered with ridges in no pattern; grey to yellowish; 2.5–2.8 mm. Seed oval, grooved. Yellow with a reddish spot by hilum. Length: 1.9–2.2 mm. Width: 1.0–1.3 mm. Map and image courtesy of USDA-NRCS.

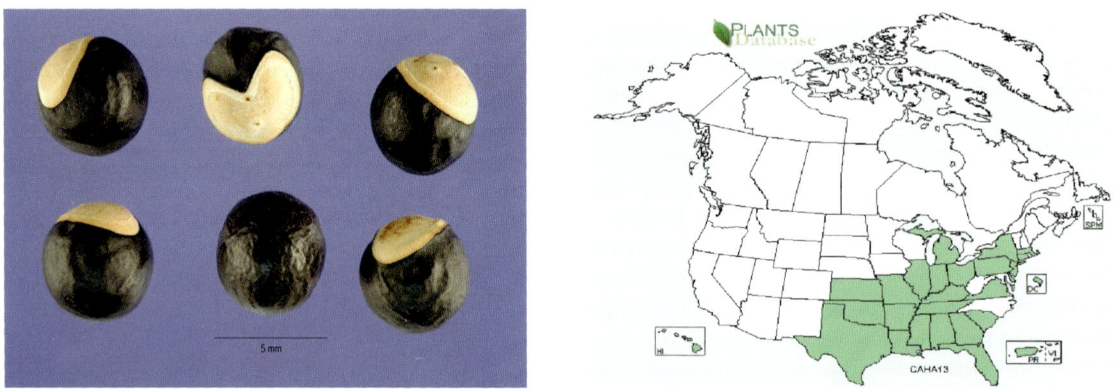

Fig. 7.2. Balloonvine (Soapberry Family) [*Cardiospermum halicacabum* (L)]. Other common names: none. Round in outline and cross section. Surface black with a white "saddle." Size: 4.5–5.0 mm. Native to tropical America. Map and image courtesy of USDA-NRCS.

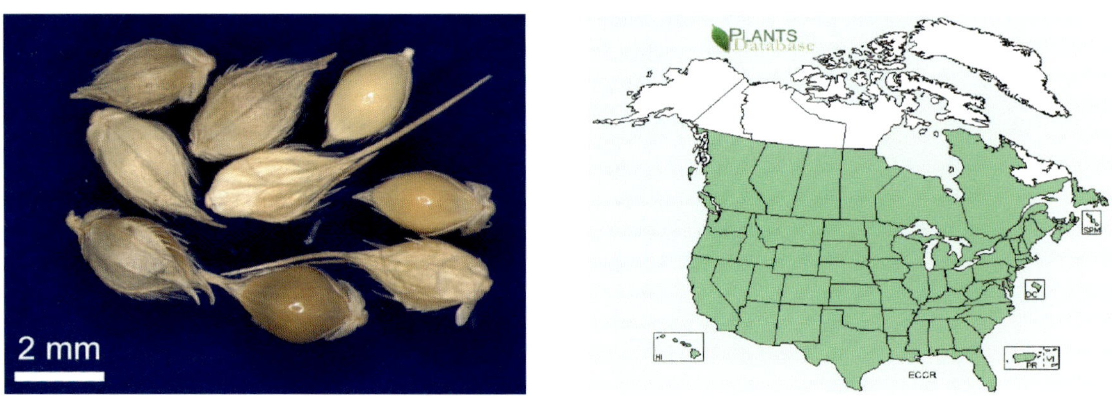

Fig. 7.3. Barnyardgrass (Grass Family) [*Echinochloa crusgalli* (L) Desv.]. Other common names: Cockspur Grass, Watergrass. Hull has a long awn and spikes; hulled seed oval in outline, plano-convex in cross section. Shiny. Buff-colored to brown. Length: 2.8–3.0 mm. Width: 1.6–2.2 mm. Introduced from Europe. © Her Majesty the Queen in Right of Canada (2010); Map courtesy of USDA-NRCS.

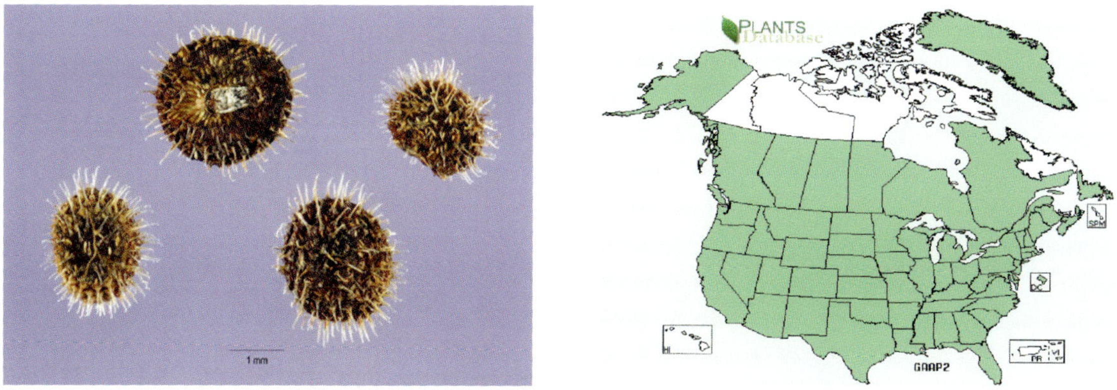

Fig. 7.4. Bedstraw (Madder Family) [*Galium aparine* (L)]. Other common names: Catchweed Bedstraw, Cleavers, Cleaverwort, Scarthgrass, White Hedge. Oval to round, spiny in appearance. Dull to shiny brown colored. Length: 2.0–3.0 mm. Width: 1.2–3.0 mm. Found in the European Union as well. Map and image courtesy of USDA-NRCS.

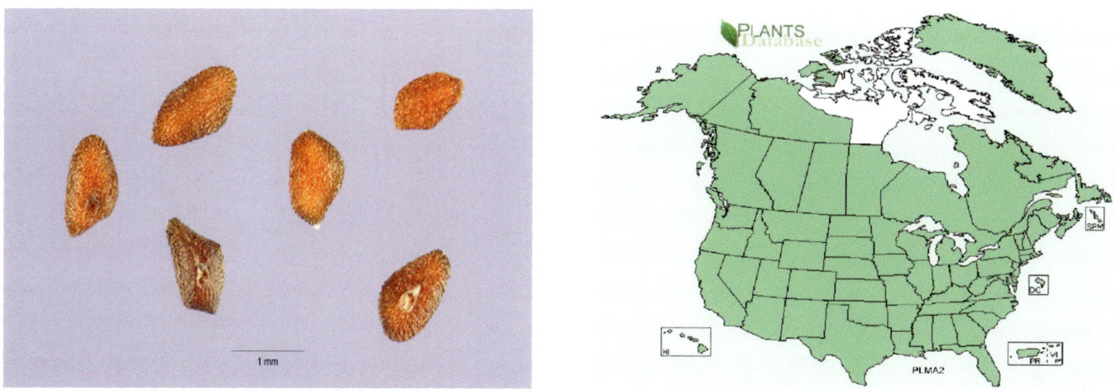

Fig. 7.5. Broadleaf Plantain (Plantain Family) [*Plantago major* (L)]. Other common names: Common Plantain, Rippleseed Plantain. Shapes variable, often triangular in outline, flattened in cross section. Light to dark brown, with a rough surface. Length: 1.0–1.5 mm. Width: 0.5–0.8 mm. Native to Eurasia. Map and image courtesy of USDA-NRCS.

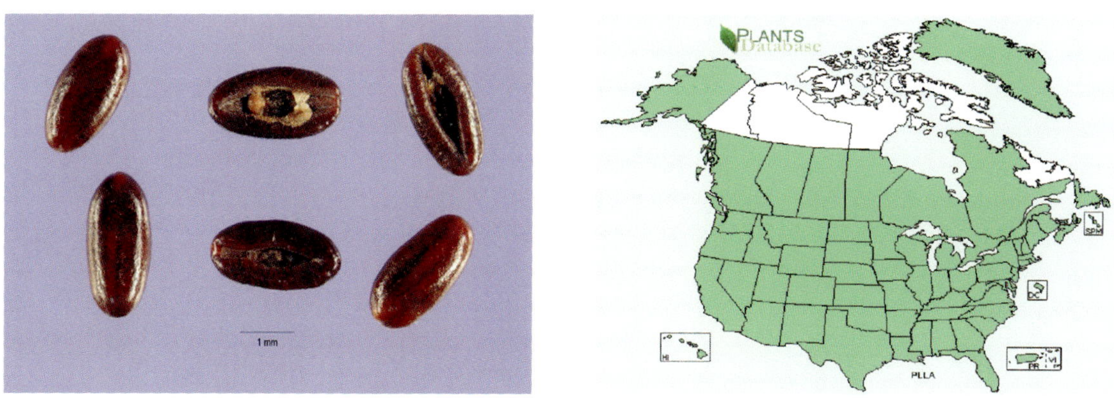

Fig. 7.6. Buckhorn Plantain (Plantain Family) [*Plantago lanceolata* (L)]. Other common names: Narrowleaf Plantain, English Plantain, Lanceleaf Plantain, Ribgrass, Ribwort. Elongated oval in outline, with a broad groove extending the entire length of the ventral side. Light to dark brown, with a shiny surface. Length: 2.3–2.7 mm. Width: 1.0–1.4 mm. Introduced from Europe. Map and image courtesy of USDA-NRCS.

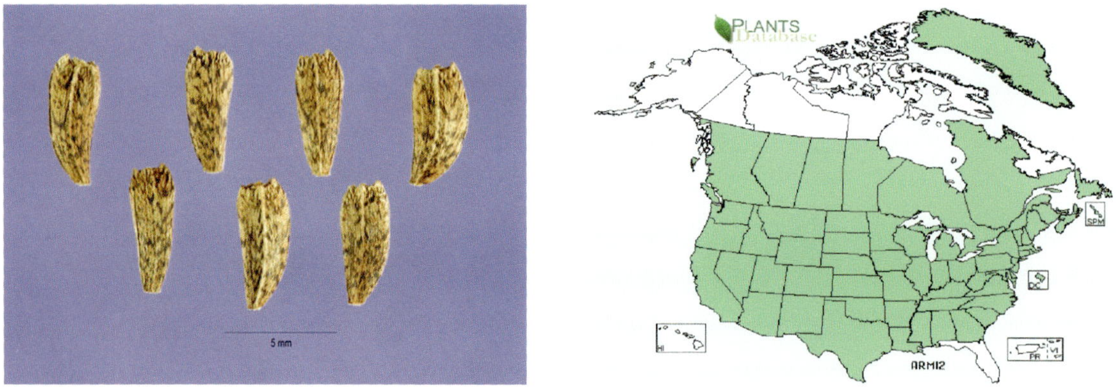

Fig. 7.7. Burdock (Aster Family) (*Arctium minus* Bernh.). Other common names: Barbane, Beggar's Button, Common Burdock, Lesser Burdock, Wild Burdock, Wild Rhubarb. Elongated oval in outline and flattened in cross section, with both ends truncated. Tan to Brown with random black markings. Surface dull and rough. Length: 4.5–6.0 mm. Width: 2.2–2.7 mm. Imported from Europe. Map and image courtesy of USDA-NRCS.

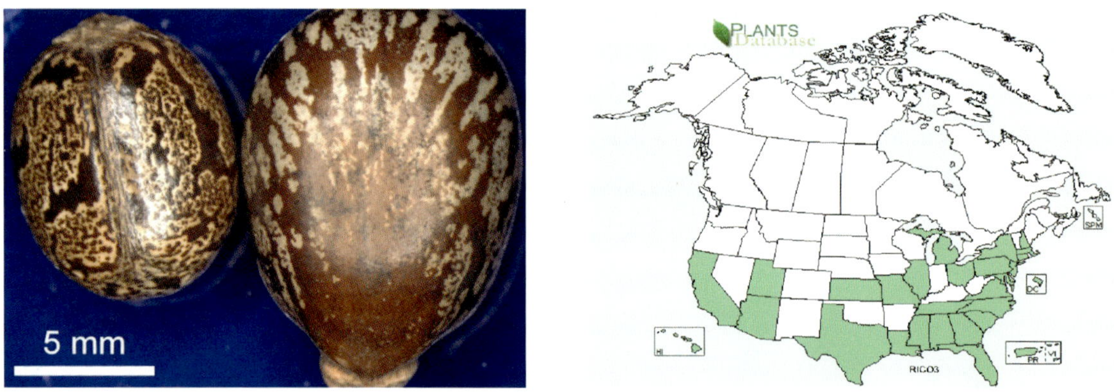

Fig. 7.8. Castorbean (Spurge Family) [*Ricinus communis* (L)]. Other common names: Castor Bean, Agaliya, Lama Palagi, Maskerekur, Toto Ni Vavalagi, Uluchula Skoki. Oval in outline, very slightly flattened in cross section. Resembles a mottled bean. Cream background with reddish brown pattern. Length: 6.5–10 mm. Width: 4.0–7.3 mm. Seed is toxic. Found in the European Union. © Her Majesty the Queen in Right of Canada (2010); Map courtesy of USDA-NRCS.

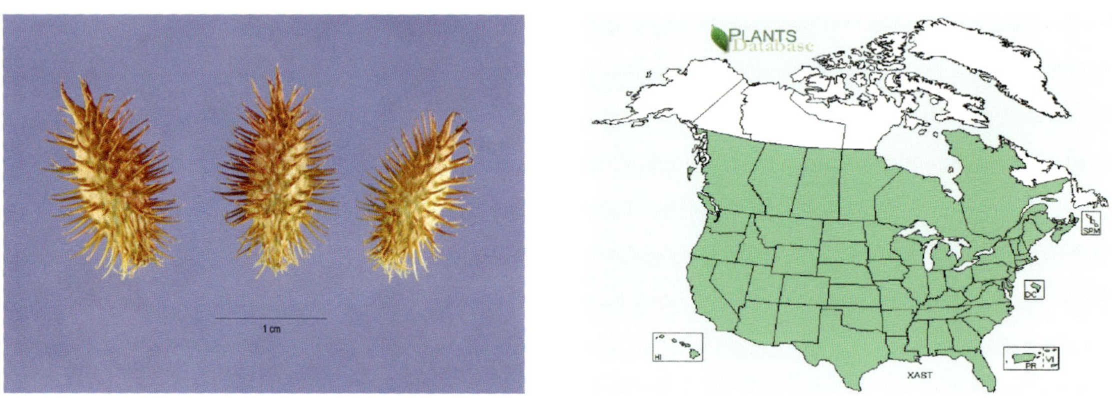

Fig. 7.9. Cocklebur (Aster Family) [*Xanthium strumarium* (L)]. Other common names: Common and Rough Cocklebur. Bur is elongated oval, round in cross section. Golden to brownish; surface covered with hooked spines. Length: 10–20 mm. Width: 10–15 mm. Seeds may be toxic. Native of Eurasia. Map and image courtesy of USDA-NRCS.

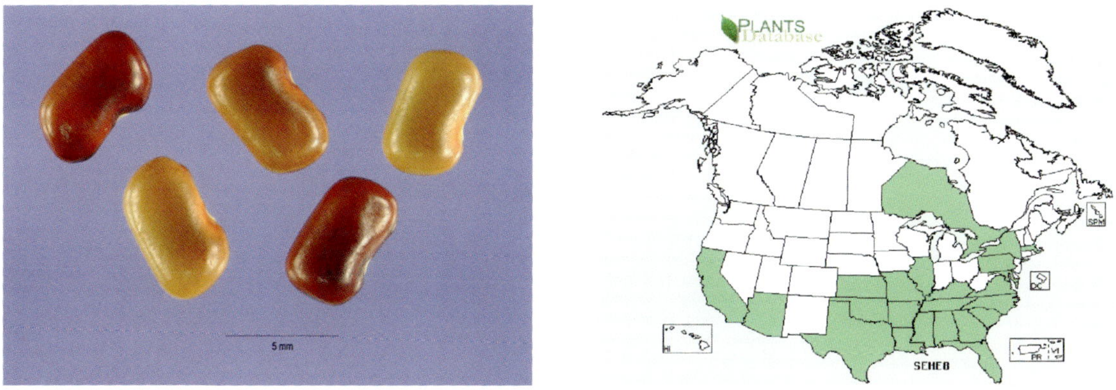

Fig. 7.10. Coffeebean (Pea Family) [*Sesbania herbacea* (Mill.) McVaugh]. Other common names: Bigpod Sesbania, Hemp Sesbania. Oblong in outline, slightly oval in cross section. Greenish to reddish brown, variously streaked and mottled with black. Length: 3.8–4.3 mm. Width: 2.0–2.5 mm. Seed may be toxic. Map and image courtesy of USDA-NRCS.

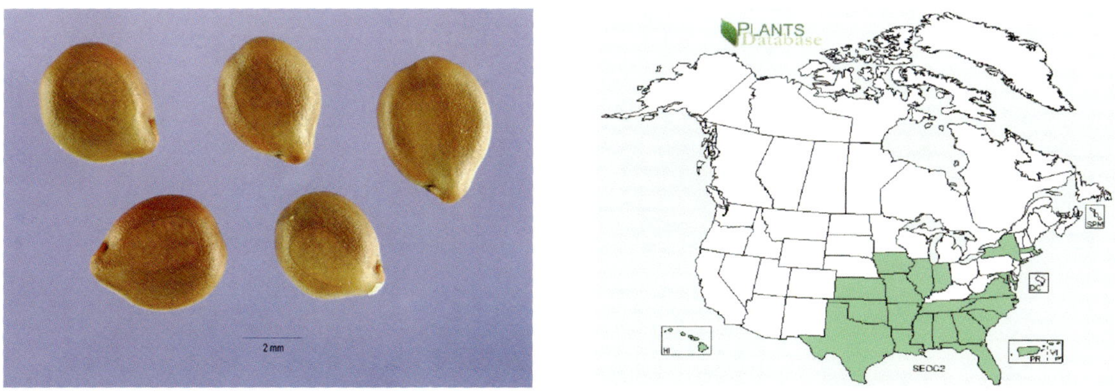

Fig. 7.11. Coffee Senna (Pea Family) [*Senna occidentalis* (L)]. Other common name: Septicweed. Oval in outline, flattened in cross section. Surface is smooth except in the center of the faces, where there is an elongated rough area. Taupe to dark brown. Length: 4.5–6.0 mm. Width: 4.0–5.0 mm. Seed may be toxic. Naturalized from the tropics. Map and image courtesy of USDA-NRCS.

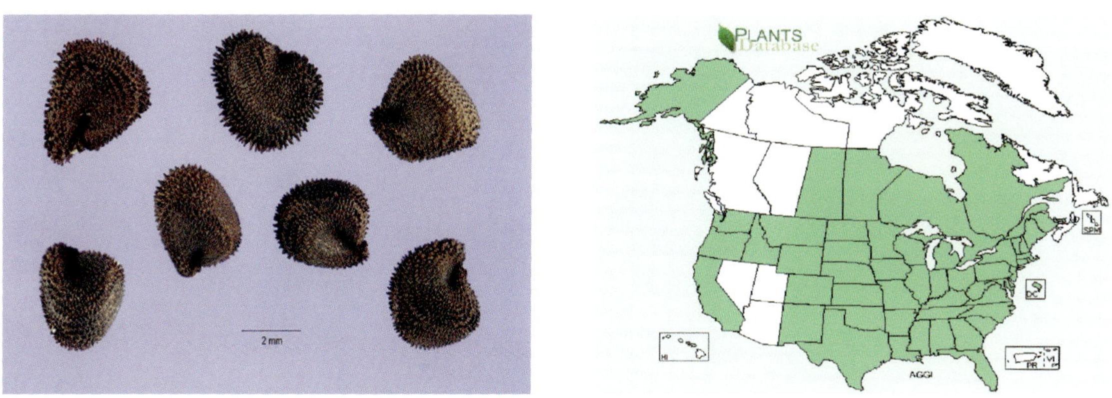

Fig. 7.12. Corncockle (Pink Family) [*Agrostemma githago* (L)]. Other common names: Corn Cockle, Common Corncockle, Purple Cockle. Triangular in outline, sector-shaped in cross section. Surface covered with "wart-like" projections. Dark brown to black. Size: 3.0–4.0 mm. Seed is toxic. Map and image courtesy of USDA-NRCS.

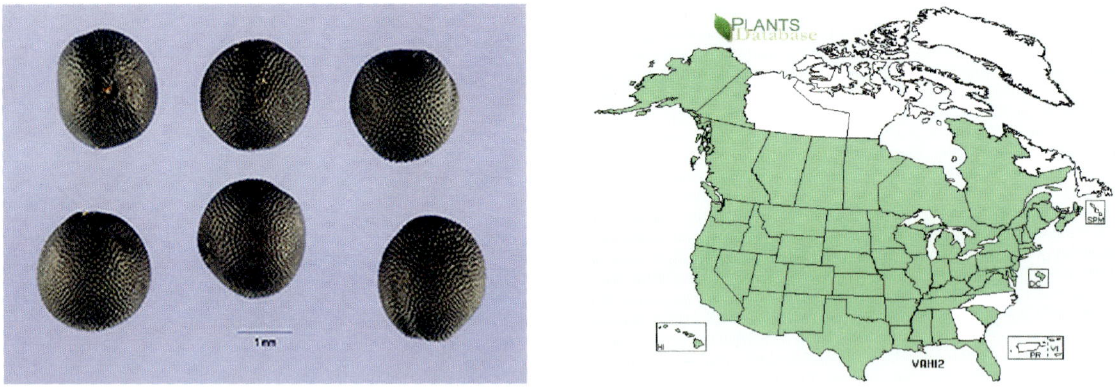

Fig. 7.13. Cow Cockle (Pink Family) [*Saponaria vaccaria* (L)]. Other common names: Cow Soapwort, Cowherb. Round with a depressed hilum. Surface covered with small projects. Reddish to black. Size: 2.0–3.0 mm. Seed may be toxic. Map and image courtesy of USDA-NRCS.

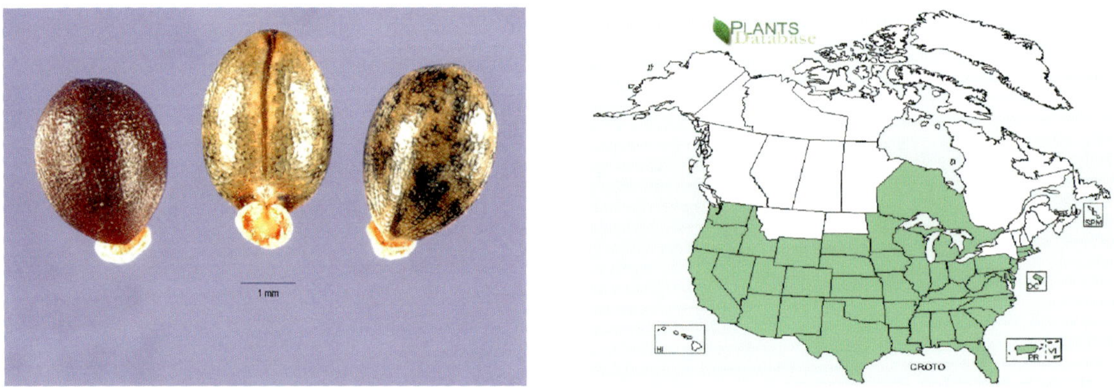

Fig. 7.14. Croton (Spurge Family) [*Croton* (L)]. Other common names: Vente Conmigo, Hogwort, Doveweed, Woolly Croton. Round to oval in shape with a prominent protruding hilum. Surface is wrinkled to dimpled in textures, often shiny with one side having a medial line splitting the seed in half starting at the hilum. Seed is solid reddish brown to tan mottled in color. Size and color varies by species. Size: 3.25–4.25 mm. Found in the European Union. Map and image courtesy of USDA-NRCS.

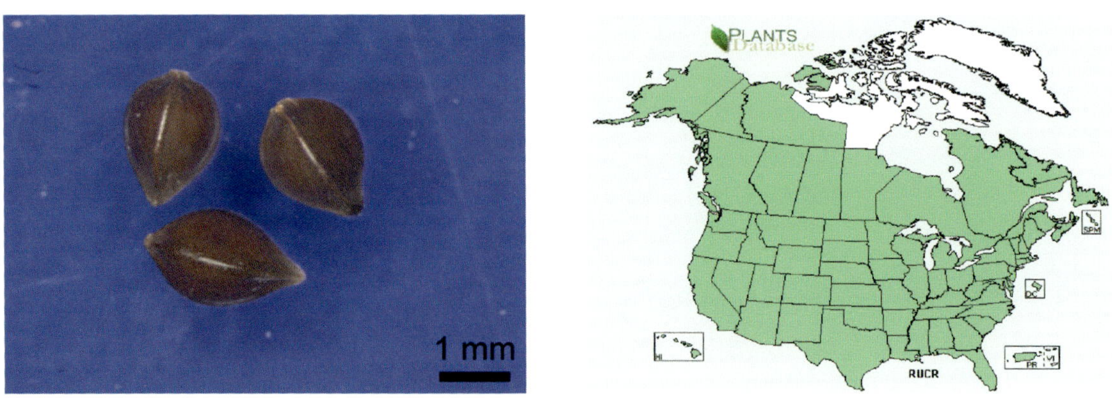

Fig. 7.15. Curly Dock (Buckwheat Family) [*Rumex crispus* (L)].Other common names: Sour Dock, Narrowleaf Dock, Yellow Dock. Oval in outline, triangular in cross section. Reddish brown, with a shiny surface. Length: 2.2–2.7 mm. Width: 1.3–1.5 mm. Naturalized from Europe. © Her Majesty the Queen in Right of Canada (2010); Map courtesy of USDA-NRCS.

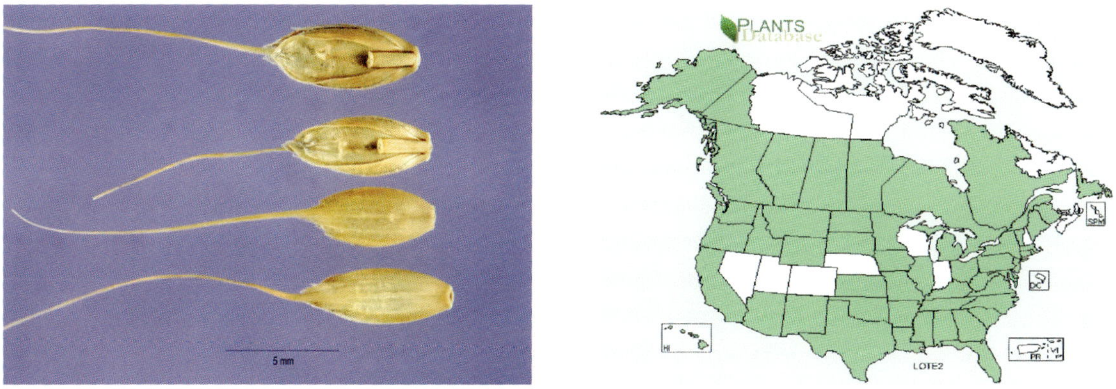

Fig. 7.16. Darnel Ryegrass (Grass Family) [*Lolium temulentum* (L)]. Other common name: Darnel. Elliptical in outline, with a groove on the ventral side. A long awn may be present. Yellowish brown to orange brown. Length: 6.0–10.0 mm. Width: 1.7–2.3 mm. Seed is toxic due to the presence of a fungus, *Endoconidium temulentum.* Introduced from Europe. Map and image courtesy of USDA-NRCS.

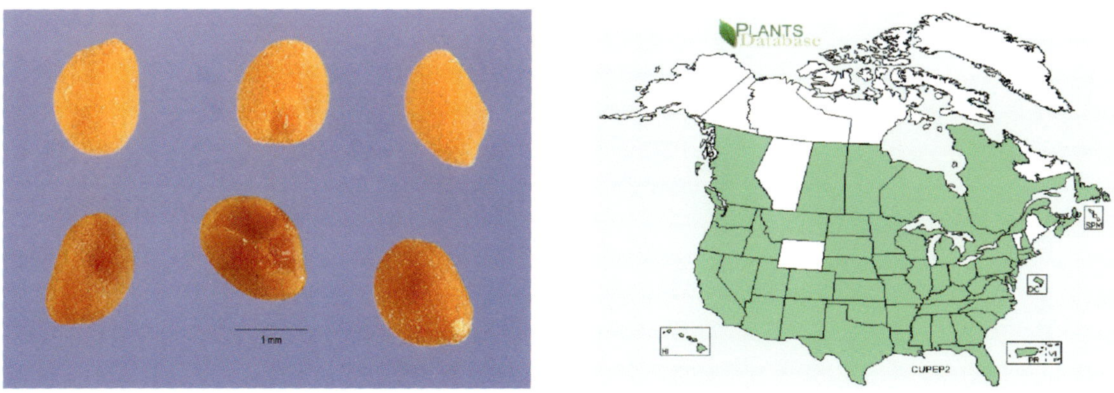

Fig. 7.17. Dodder (Dodder Family) (*Cuscuta sp.*). Other common names: *Cuscuta pentagona,* Fiveangled Dodder. Oval in outline, triangular in cross section. Light brown to yellowish brown to golden, with a roughened surface. Size: 1.3–1.5 mm. Native to North America. Map and image courtesy of USDA-NRCS.

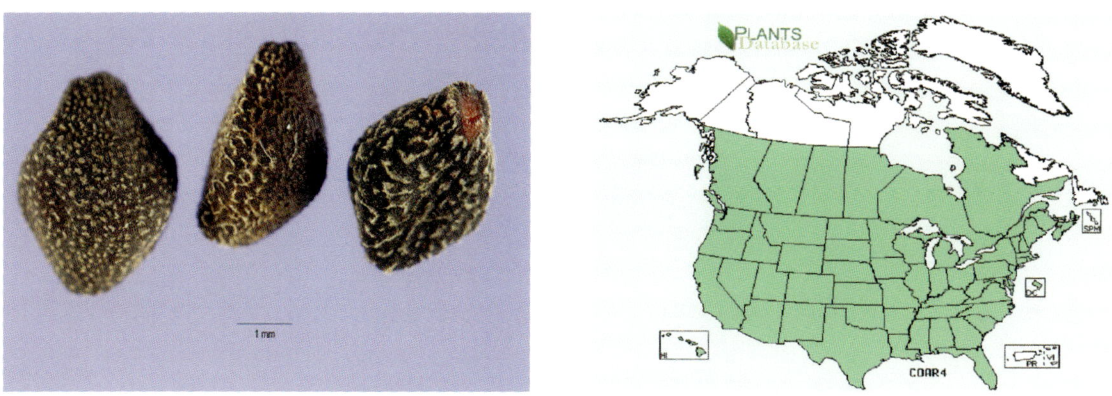

Fig. 7.18. Field Bindweed (Morning-glory Family) [*Convolvulus arvensis* (L)]. Other common names: Creeping Jenny, Smallflowered Morning-glory, Perennial Morning-glory, European Bindweed. Broadly oval in outline, sector-shaped in cross section. Grayish brown, with a roughened surface. Seed scar has knob-like projections on either side. Length: 4.0–4.5 mm. Width: 2.5–3.0 mm. Introduced from Eurasia. Map and image courtesy of USDA-NRCS.

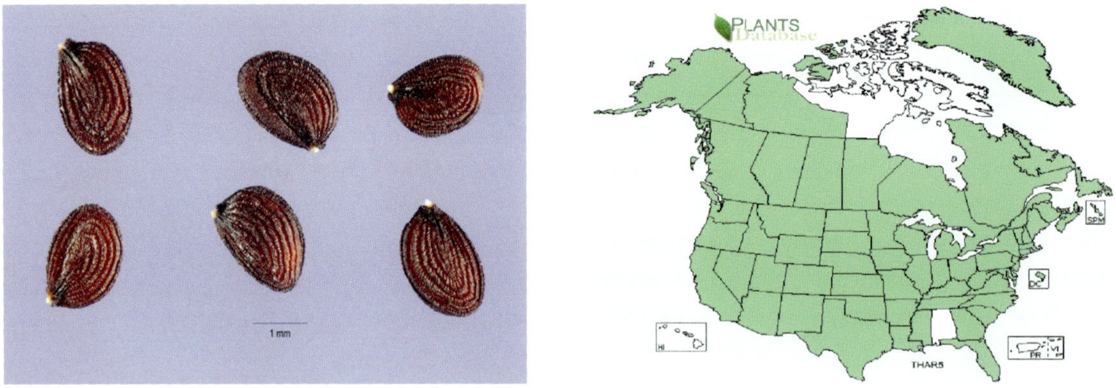

Fig. 7.19. Field Pennycress (Mustard Family) [*Thlaspi arvense* (L)]. Other common names: Pennycress, Fanweed, Frenchweed, Stinkweed. Oval in outline with slightly elongated base. Surface roughened by concentric ridges. Flattened in cross section. Dark brown to black. Length: 2.0–2.4 mm. Width: 1.1–1.4 mm. Seed may be toxic. Native to Eurasia. Map and image courtesy of USDA-NRCS.

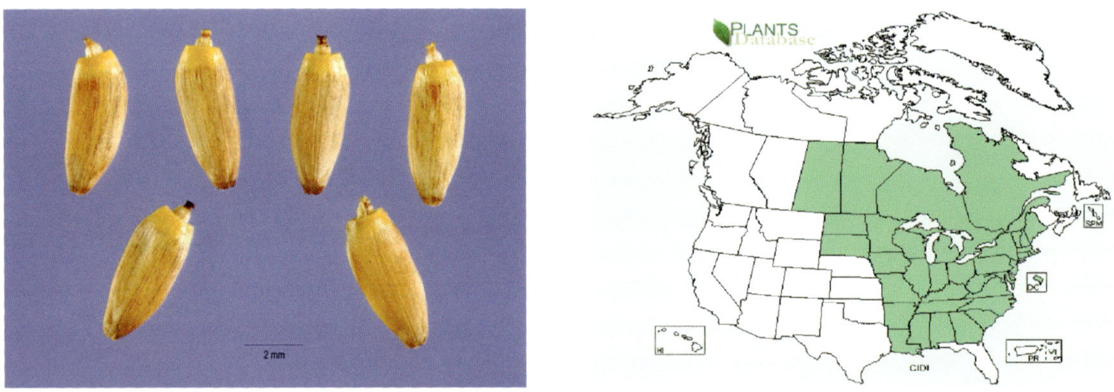

Fig. 7.20. Field Thistle (Aster Family) [*Cirsium discolor* (Muhl. Ex Willd.) Spreng.]. Other common names: Thistle, Roadside Thistle. Elongated in outline, truncated on one end and pointed on the other. Flattened in cross section. Light yellowish with streaks of purple or black; a bright yellow band circles the truncated end. Length: 4.1–5.2 mm. Width: 1.8–2.4 mm. Map and image courtesy of USDA-NRCS.

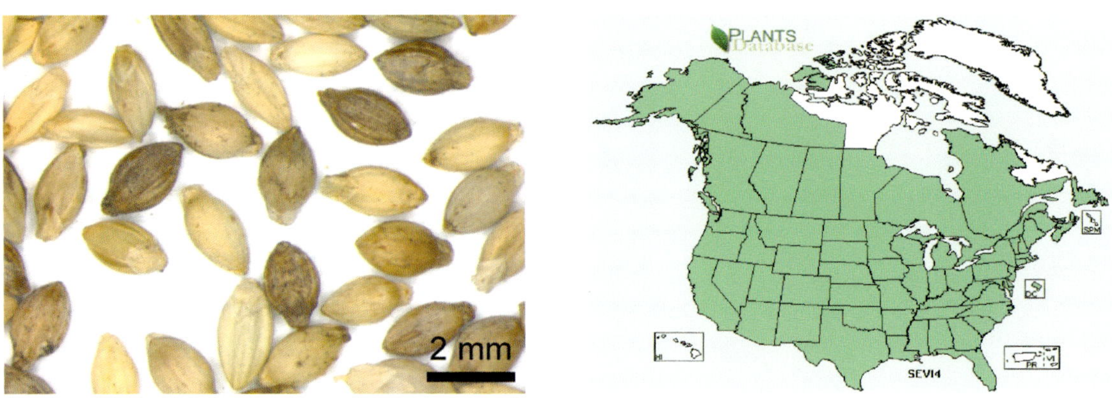

Fig. 7.21. (Green) Foxtail (Grass Family) [*Setaria viridis* (L) P. Beauv.]. Other common names: Bottlegrass, Green Bristlegrass, Green Millet, Pigeongrass. Oval in outline, plano-convex in cross section. Surface roughened with transverse ridges. Yellow to dark brown. Length: 1.8–2.2 mm. Width: 1.0–1.3 mm. Native to Europe. © Her Majesty the Queen in Right of Canada (2010); Map courtesy of USDA-NRCS.

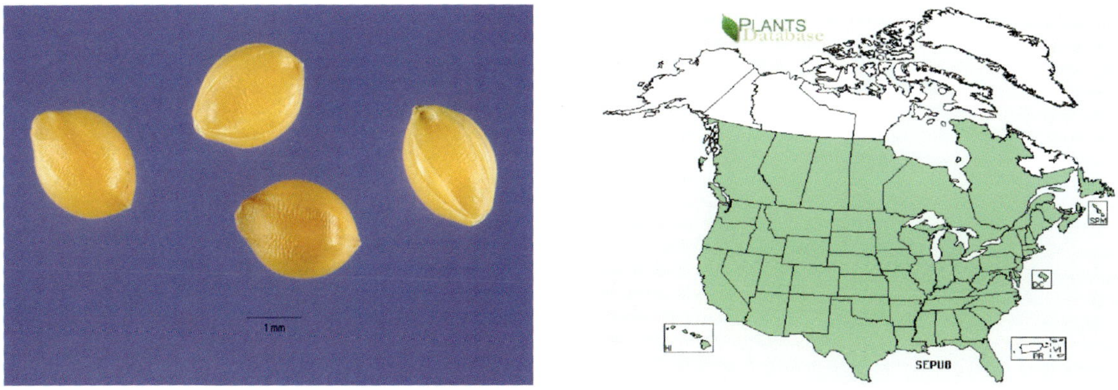

Fig. 7.22. (Yellow) Foxtail (Grass Family) [*Setaria pumila* (Poir.) Roem. & Schult]. Other common names: Cattail Grass, Smooth Millet, Pigeongrass, Yellow Bristlegrass. Oval in outline, plano-convex in cross section. Surface roughened with transverse ridges. Greenish, golden yellow or dark brown. Length: 2.5–2.8 mm. Width: 1.8–2.1 mm. Naturalized from Europe. Map and image courtesy of USDA-NRCS.

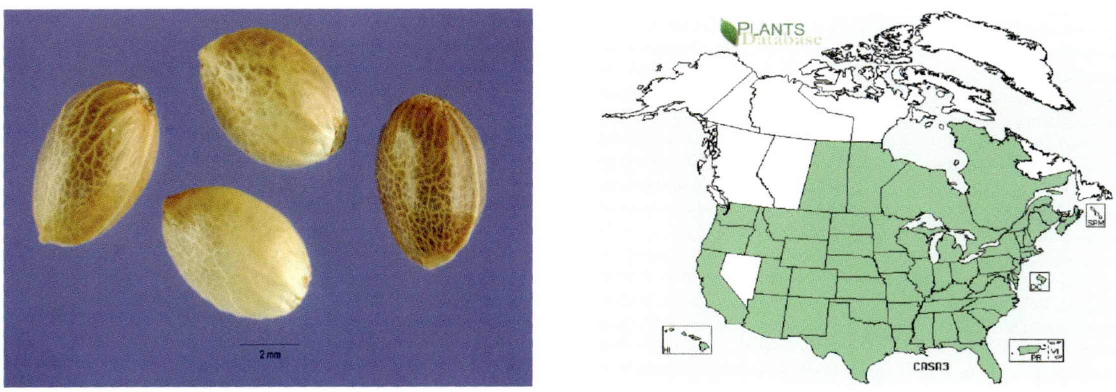

Fig. 7.23. Hemp (Hemp Family) [*Cannabis sativa* (L)]. Other common names: Marijuana, Mary Jane, Hashish, Pot, Grass. Oval in outline and cross section. Taupe through olive green to dark brown, with a network of lines on the surface. Length: 3.5–5.0 mm. Width: 2.0–3.0 mm. Seed is toxic; possession is illegal unless seed is sterilized, or by license. Native of Asia. Map and image courtesy of USDA-NRCS.

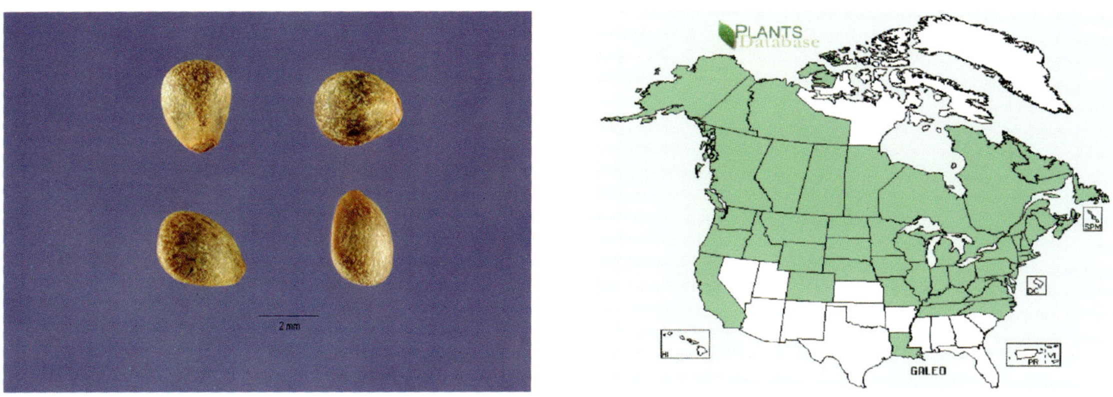

Fig. 7.24. Hempnettle (Mint Family) [*Galeopsis tetrahit* (L)]. Other common name: none. Broadly oval in outline, sector-shaped in cross section. Grayish brown, surface darkly speckled and slightly rough. Length: 2.5–3.3 mm. Width: 1.5–2.0 mm. Introduced from Eurasia. Map and image courtesy of USDA-NRCS.

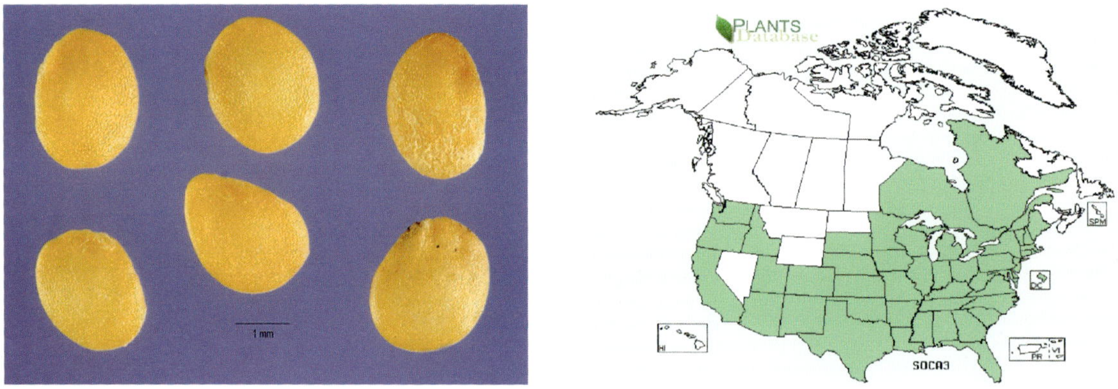

Fig. 7.25. Horsenettle (Potato Family) [*Solanum carolinense* (L)]. Other common names: Carolina Nettle, Bull Nettle, Bull Thistle. Circular in outline, flattened in cross section. Surface irregularly pebbled, often covered with debris due to its sticky nature. Yellowish to golden brown. Size: 3.0 mm. Seed thought to be toxic. Map and image courtesy of USDA-NRCS.

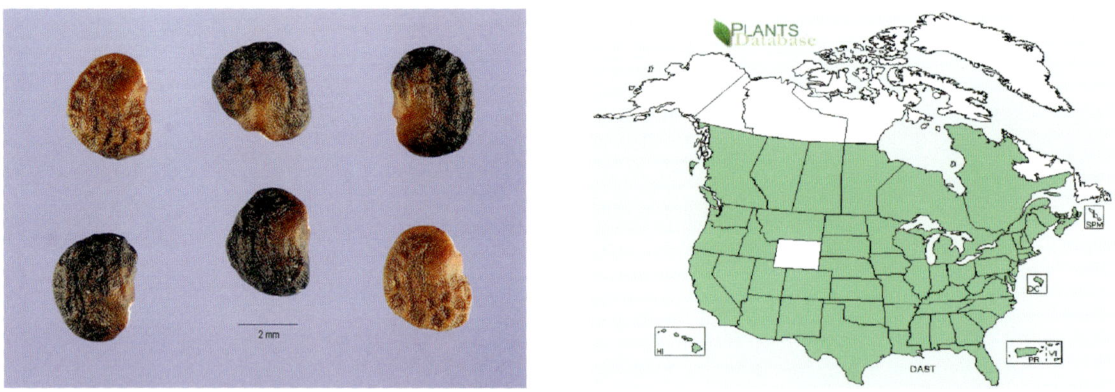

Fig. 7.26. Jimsonweed (Potato Family) [*Datura stramonium* (L)]. Other common names: Thorn Apple, Jamestown Weed, Mad Apple, Moonflower, Stinkwort. Slightly kidney-shaped in outline, flattened in cross section. Surface bears a network of thick veins. Hilum is triangular, lying on the side. Buff-colored to black. Length: 3.0–3.8 mm. Width: 2.5–3.0 mm. Seed is toxic. Found in the European Union. Introduced from Africa, native to India. Map and image courtesy of USDA-NRCS.

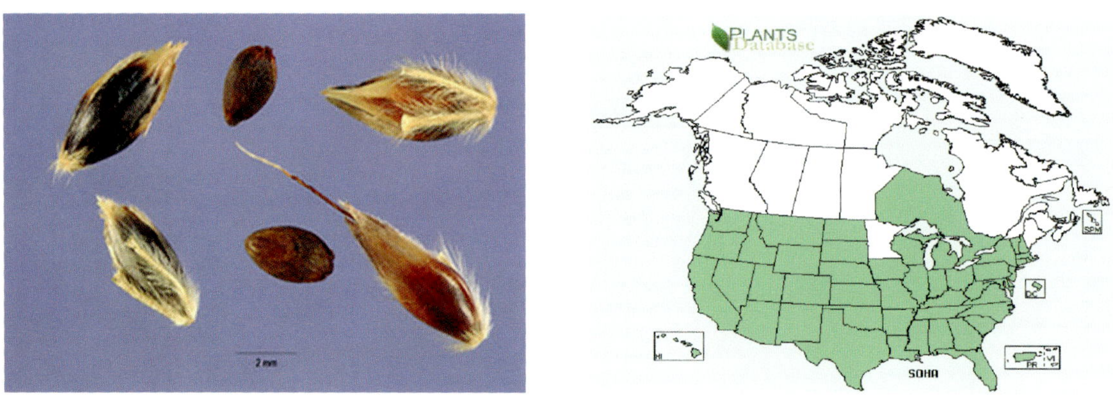

Fig. 7.27. Johnsongrass (Grass Family) [*Sorghum halepense* (L) Pers.]. Other common names: Aleppo Milletgrass, Herbe de Cuba, Sorgho de Alep, Zacate Johnson, Columbus Grass. Elliptical in outline, plano-convex in cross section. Buff through red to black. Often has trumpet-shaped rachillae. Length: 2.0–3.0 mm. Width 1.5–2.0 mm. Map and image courtesy of USDA-NRCS.

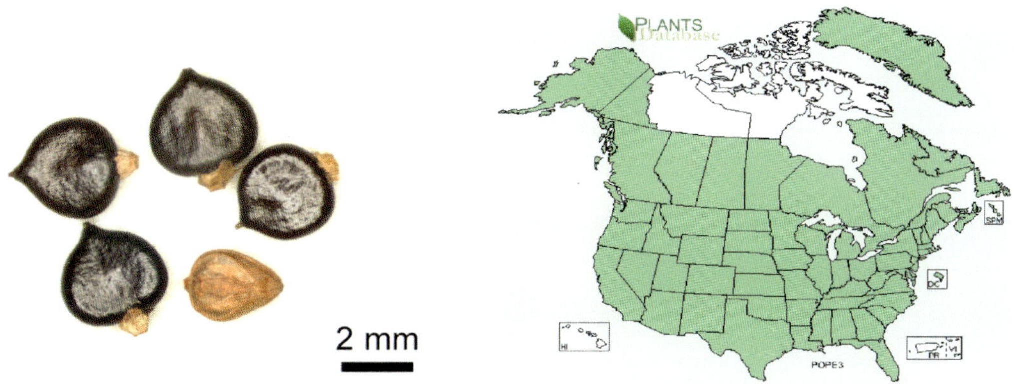

Fig. 7.28. Ladysthumb (Buckwheat Family) [*Polygonum persicaria* (L)]. Other common names: Lady's Thumb, Smartweed, Ladysthumb Smartweed. Heart-shaped in outline, flattened in cross section. Reddish brown to dark brown, with a shiny surface. Length: 2.0–2.8 mm. Width: 1.5–1.9 mm. Introduced from Europe. © Her Majesty the Queen in Right of Canada (2010); Map courtesy of USDA-NRCS.

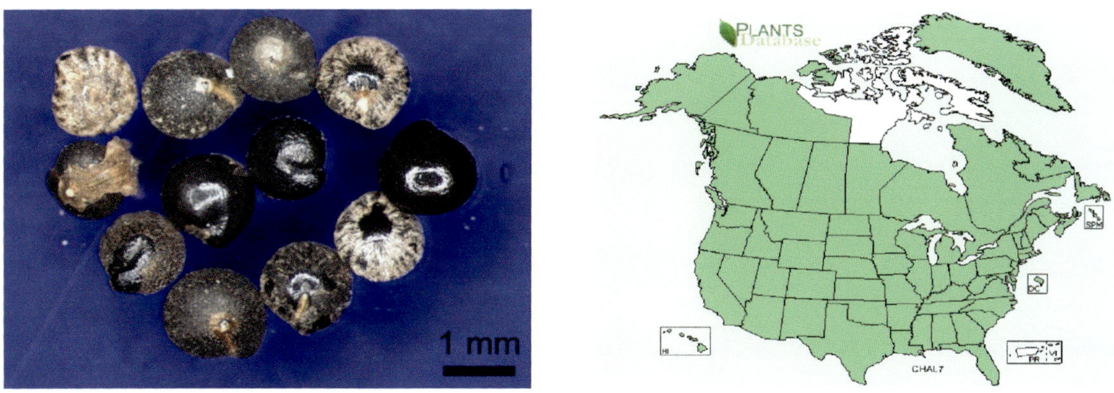

Fig. 7.29. Lambsquarters (Goosefoot Family) [*Chenopodium album* (L)]. Other common names: White Goosefoot, Common Lambsquarters. Generally round in outline, lens-shaped in cross section, with one side flattened. Margin notched at hilum, with a groove running inward. Black. Size: 1.2–1.6 mm. Native to Eurasia. © Her Majesty the Queen in Right of Canada (2010); Map courtesy of USDA-NRCS.

Fig. 7.30. Milkweed (Milkweed Family) [*Asclepias syriaca* (L)]. Other common name: Common Milkweed. Oval in outline and flattened in cross section, with a truncated tip. Edges flanged. Brown, with a rough surface. Length: 8.0–10.5 mm. Width: 4.5–5.5 mm. Seed may be toxic. Native to North America. Map and image courtesy of USDA-NRCS.

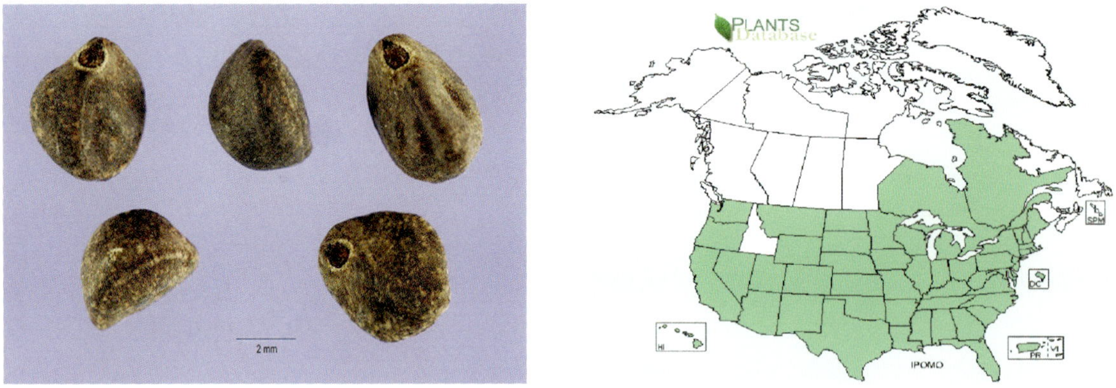

Fig. 7.31. Morning-Glory (Morning-glory Family) [*Ipomoea* (L)]. Other common name: Morning Glory. Shapes variable. Brown to black; some surfaced smooth, and others roughened. Size: 4.5–7.0 mm. Hilum generally prominent. Naturalized from the American tropics. Map and image courtesy of USDA-NRCS.

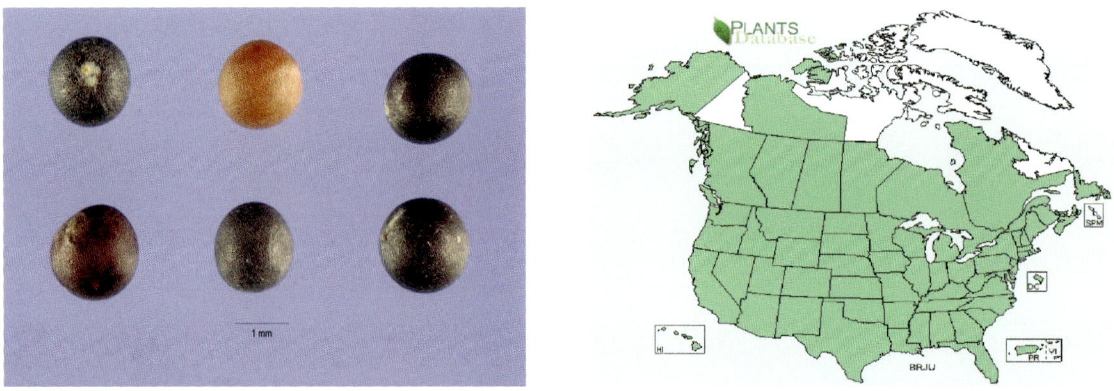

Fig. 7.32. Mustard (Mustard Family) [*Sinapis arvensis* (L)]. Other common names: Wild Mustard, Charlock Mustard, Canola, Kaber Mustard, Rapeseed. Generally round. Surface is granular. Yellow to black or dark maroon, immature seeds are red. Size: 1.0–2.0 mm. Seed may be toxic. Found in the European Union. Introduced from Europe. Map and image courtesy of USDA-NRCS.

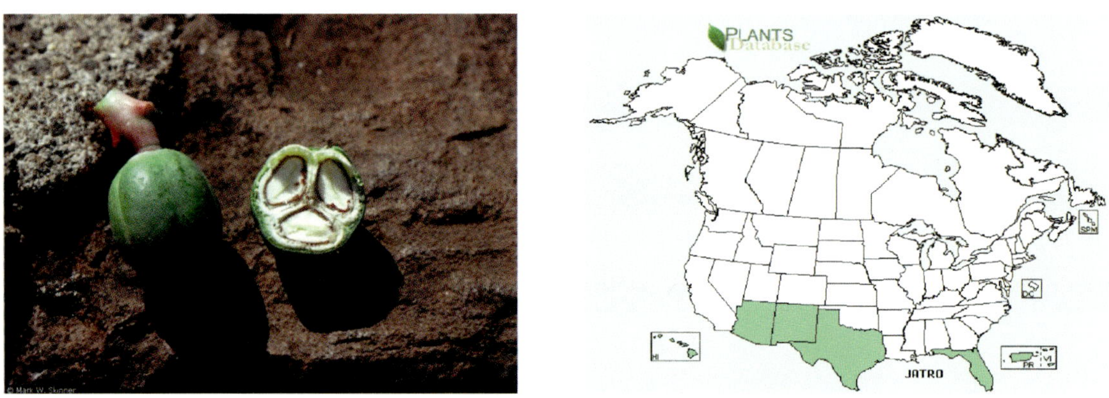

Fig. 7.33. Nettlespurge (Spurge Family) [*Jatropha* (L)]. Other common names: Barbados Nut, Purghera, Leatherstem, Purging Nut. Green to yellow colored fruiting body (shown), when split reveals three triangular black seeds. Length: 2.5 cm. Scale varies by species within genus. Found in the European Union. © Mark W. Skinner; Map courtesy of USDA-NRCS.

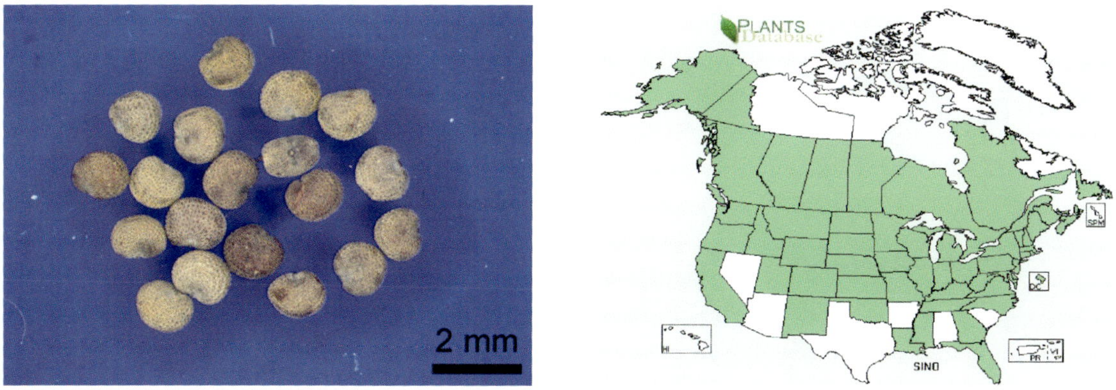

Fig. 7.34. Nightflowering Catchfly (Pink Family) [*Silene noctiflora* (L)]. Other common name: Sticky Cockle. Kidney-shaped in outline, slightly convex in cross section. Surface covered with small projections. Yellowish tan to grey to grey-brown. Length: 1.0–1.5 mm. Width 0.8–1.2 mm. © Her Majesty the Queen in Right of Canada (2010); Map courtesy of USDA-NRCS.

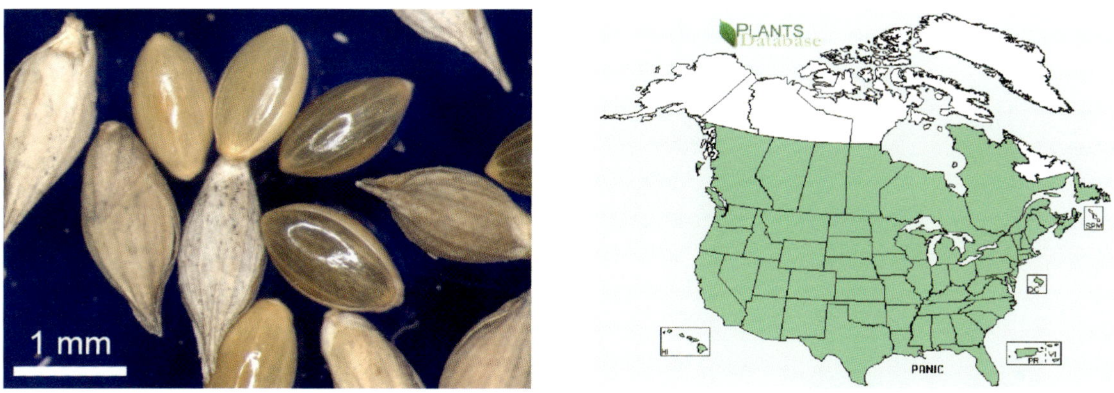

Fig. 7.35. Panicum (Grass Family) [*Panicum* (L)]. Other common name: Panicgrass, Knee-grass, Sprouting Crabgrass. Elliptical in outline, plano-convex in cross section. Greenish yellow to olive green, surface covered with yellow, longitudinal stripes. Length: 1.1–1.6 mm. Width: 0.5–0.7 mm. © Her Majesty the Queen in Right of Canada (2010); Map courtesy of USDA-NRCS.

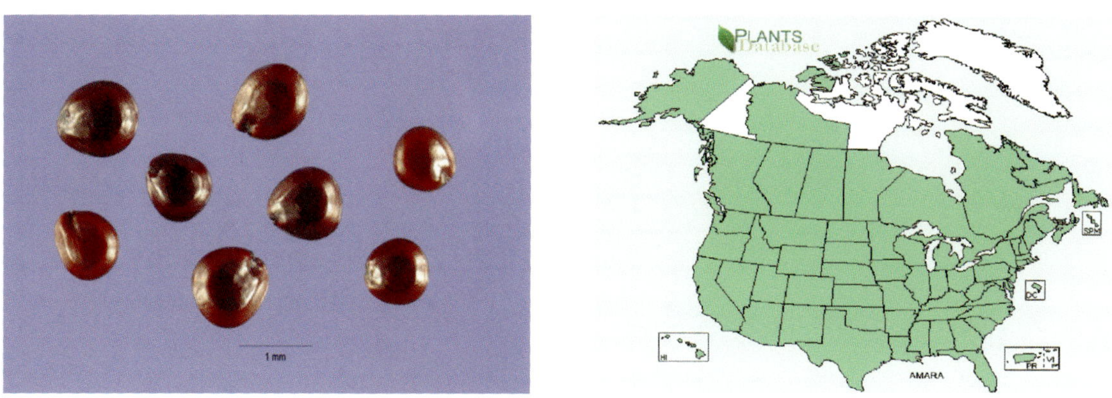

Fig. 7.36. Pigweed (Amaranth Family) [*Amaranthus* (L)]. Other common names: Carelessweed. Circular in outline, lens-shaped in cross section. Edges slightly flanged. Reddish to black, with a glassy surface. Size: 1.0–1.5 mm. Imported from tropical Americas. Map and image courtesy of USDA-NRCS.

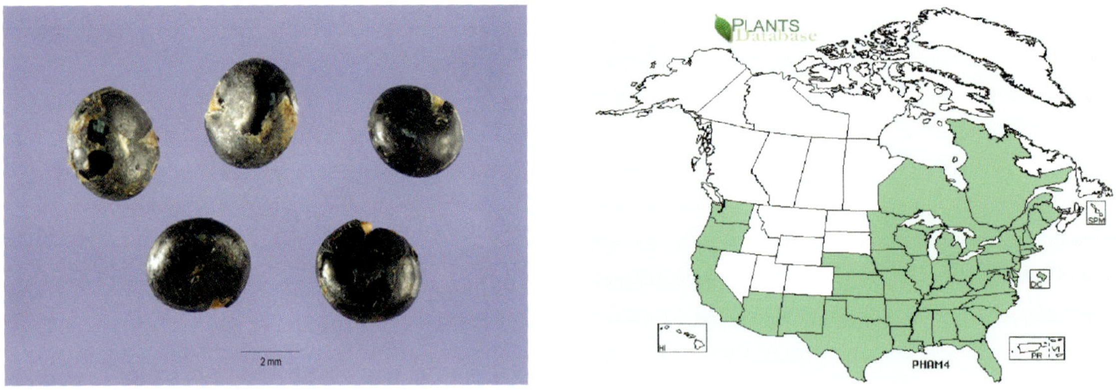

Fig. 7.37. Pokeweed (Pokeweed Family) [*Phytolacca Americana* (L)]. Other common names: American Pokeweed, Common Pokeweed, Pokeberry, Pigeonberry, Inkberry. Oval in outline and lens-shaped in cross section, with a notch on one side. Black with a shiny surface. Length: 4.0–5.0 mm. Width: 3.0–4.0 mm. Seed is toxic. Native to North America. Map and image courtesy of USDA-NRCS.

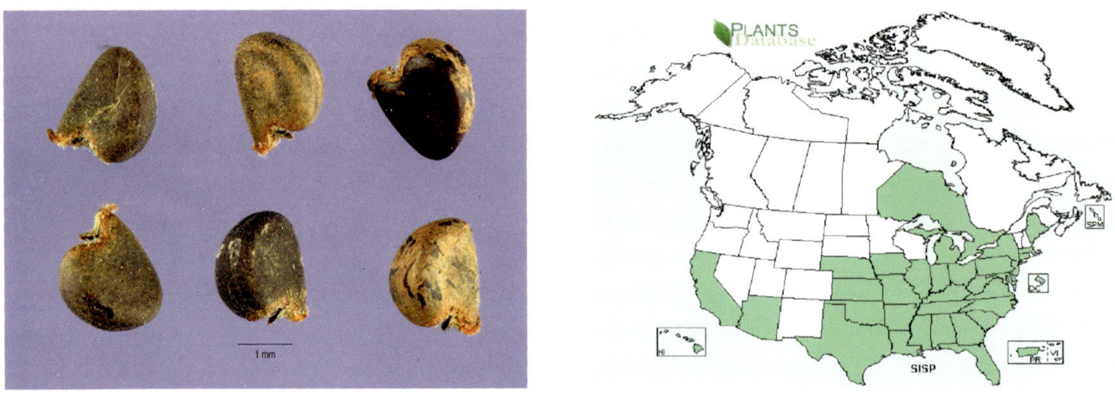

Fig. 7.38. Prickly Sida (Mallow Family) [*Sida spinosa* (L)]. Other common name: Prickly Fan-petals. Coat with two spiny projections, surface has a lacy appearance. Seed sector-shaped. Brown. Coat length: 4.0–5.0 mm. Coat width: 2.5–3.0 mm. Seed length: 2.2–2.8 mm. Seed width: 1.6–2.1 mm. Naturalized from the Tropics. Map and image courtesy of USDA-NRCS.

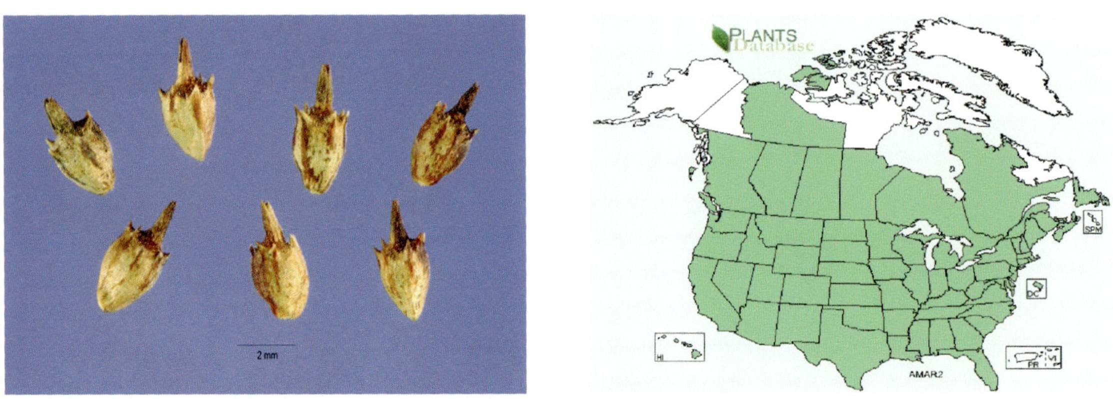

Fig. 7.39. Ragweed (Aster Family) [*Ambrosia artemisiifolia* (L)]. Other common names: Common Ragweed, Annual Ragweed, Low Ragweed, Roman Wormwood. Urn-shaped in outline, with long central protuberance on end surrounded by spines. Oval in cross section. Light brown to dark brown, with a wrinkled surface that appears woody. Length: 3.0–4.0 mm. Width: 1.8–2.5 mm. Found in the European Union. Native to Europe. Map and image courtesy of USDA-NRCS.

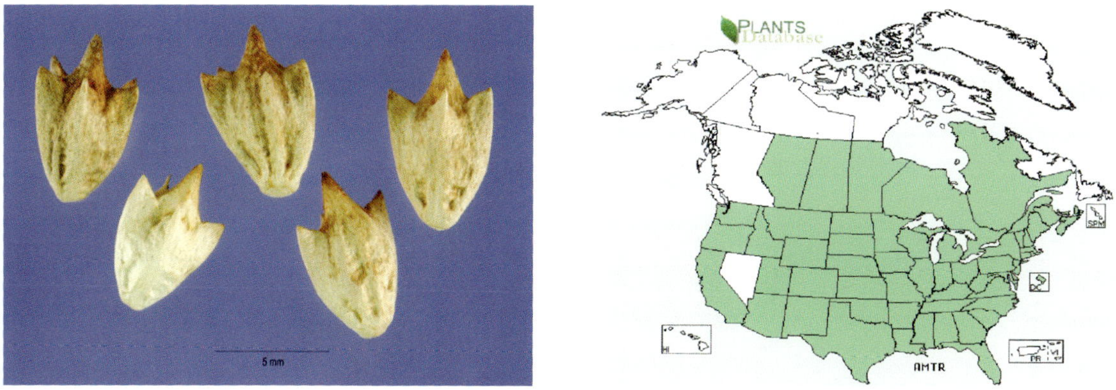

Fig. 7.40. (Giant) Ragweed (Aster Family) [*Ambrosia trifida* (L)]. Other common names: Horseweed, Great Ragweed. Roughly triangular in outline, with spines on the wider end. Oval in cross section. Light brown to gray, with a wrinkled surface that appears woody. Length: 6.0–13.0 mm. Width: 4.0–7.0 mm. Native to Europe. Map and image courtesy of USDA-NRCS.

Fig. 7.41. Saint Johnswort (Mangosteen Family) [*Hypericum perforatum* (L)]. Other common names: Klamathweed, St. John's Wort, Common St. John's Wort. Oval to rectangular in shape. Seed is reddish brown to brown with the surface raised and appears to be in a honeycomb pattern. Length: 0.5–0.75 mm. Width: 0.25–0.5 mm. Found in the European Union. Map and image courtesy of USDA-NRCS.

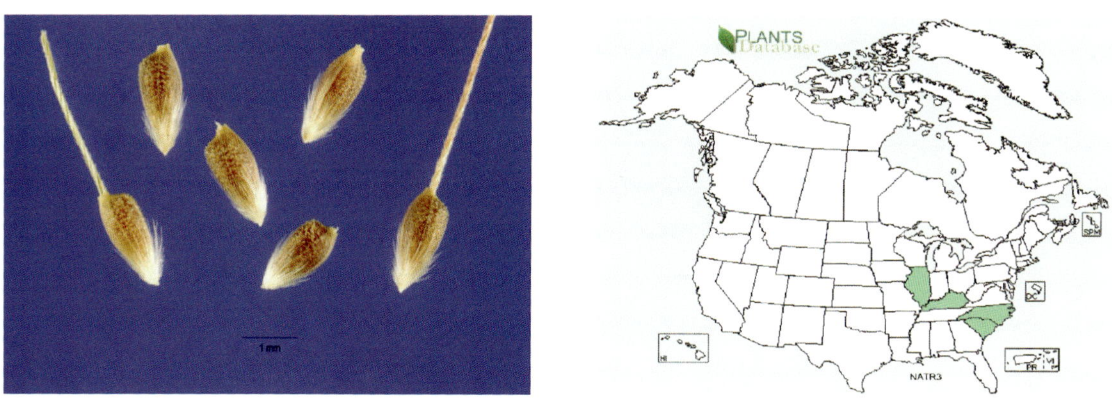

Fig. 7.42. Serrated Tussock (Grass Family) [*Nassella trichotoma* (Nees) Hack]. Other common names: Serrated Tussock Grass, Nassella tussock, Stipa trichotoma. Irregular triangular in shape, bumpy texture with a pointed tufted end. Brown to light tan. When present, awns are twisted and up to 2.5 cm. Length: 1.75–2.0 mm. Width: 0.50–0.75 mm. Map and image courtesy of USDA-NRCS.

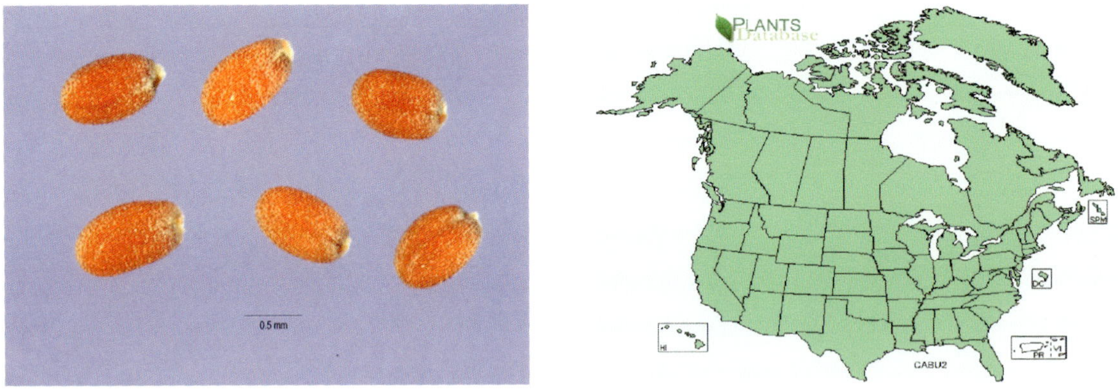

Fig. 7.43. Shepherd's Purse (Mustard Family) [*Capsella bursa-pastoris* (L) Medik]. Other common name: Shepherdspurse. Elongated oval in outline, flattened in cross section. A groove bisects each side. Yellowish brown to golden. Length: 9.9–1.2 mm. Width: 0.4–0.7 mm. Native to southern Europe. Map and image courtesy of USDA-NRCS.

Fig. 7.44. Showy Crotalaria (Pea Family) (*Crotalaria spectabilis* Roth). Other common names: Rattlebox, Showy Rattlebox. Mitten-shaped in outline, somewhat flattened in cross section, with a deep notch. Greenish brown to black, with a shiny surface. Length: 4.5–5.0 mm. Width: 3.5–4.0 mm. Seed is toxic. Map and image courtesy of USDA-NRCS.

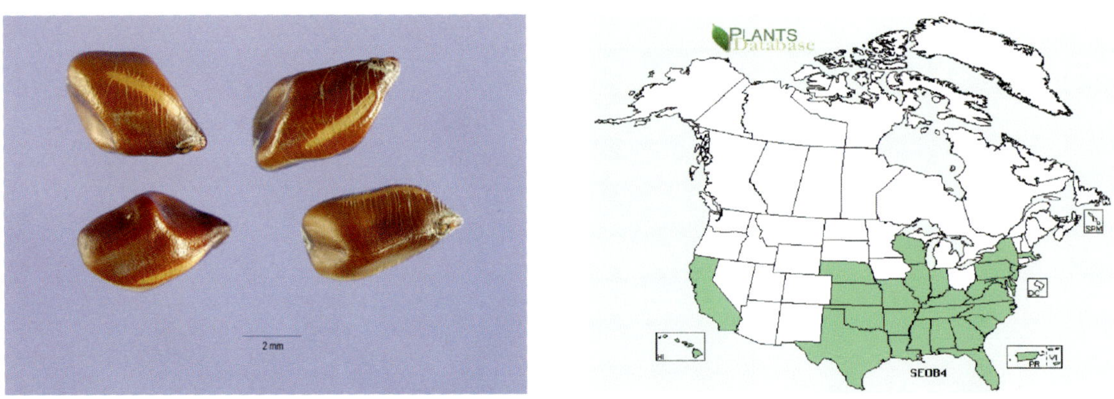

Fig. 7.45. Sicklepod (Pea Family) [*Senna obtusifolia* (L) Irwin & Barneby]. Other common names: Sicklepod Senna, Cassia obtusifolia, Java Bean. Rectangular, with a broad yellow stripe lying diagonally across the seed. Chocolate brown. Length: 5.0–6.0 mm. Width: 2.5–3.0 mm. Seed may be toxic. Map and image courtesy of USDA-NRCS.

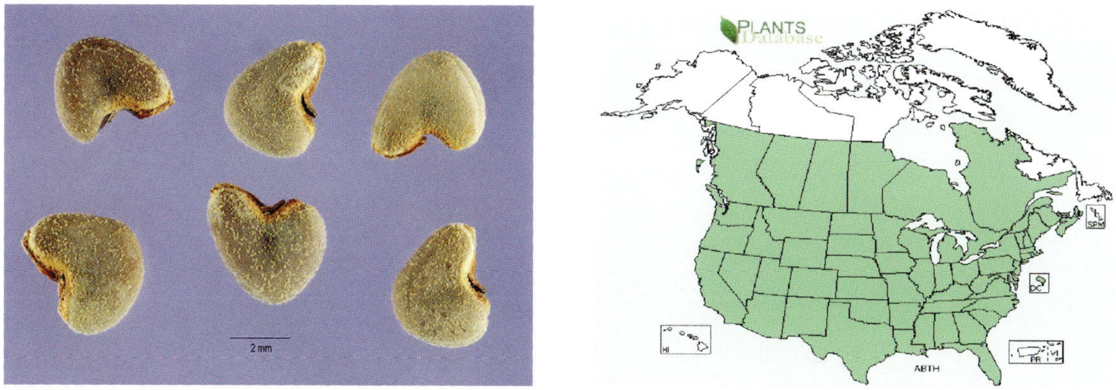

Fig. 7.46. Velvetleaf (Mallow Family) (*Abutilon theophrasti* Medik). Other common names: Buttonweed, Velvetleaf, Butterprint, Indian Mallow. Mitten-shaped in outline, somewhat concave in cross section. Coffee brown to gray-black. Length: 3.0–3.5 mm. Width: 2.2–2.6 mm. Naturalized from India. Map and image courtesy of USDA-NRCS.

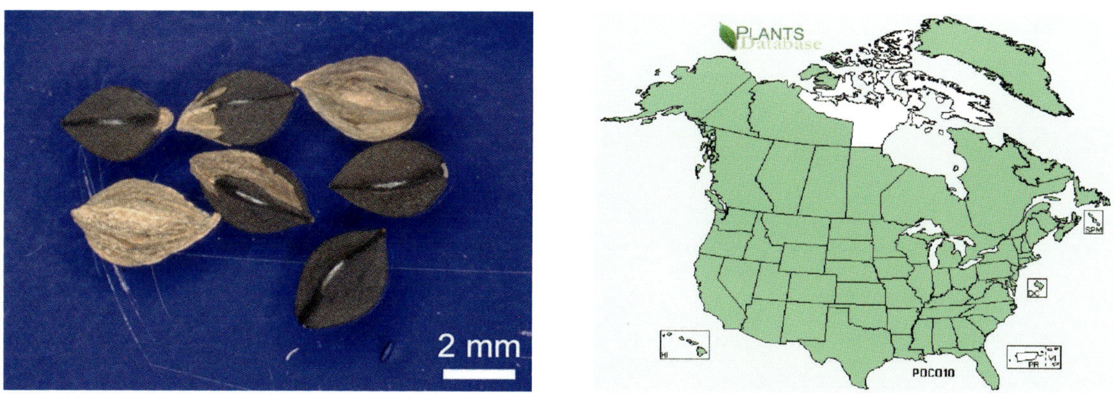

Fig. 7.47. Wild Buckwheat (Buckwheat Family) [*Polygonum convolvulus* (L)]. Other common names: Black Bindweed, Climbing Buckwheat, Cornbind, Dullseed Cornbind. Oval in outline, triangular in cross section. Dark brown to black, with a rough surface. Length: 3.0–4.0 mm. Width: 2.0–2.5 mm. Naturalized from Eurasia. © Her Majesty the Queen in Right of Canada (2010); Map courtesy of USDA-NRCS.

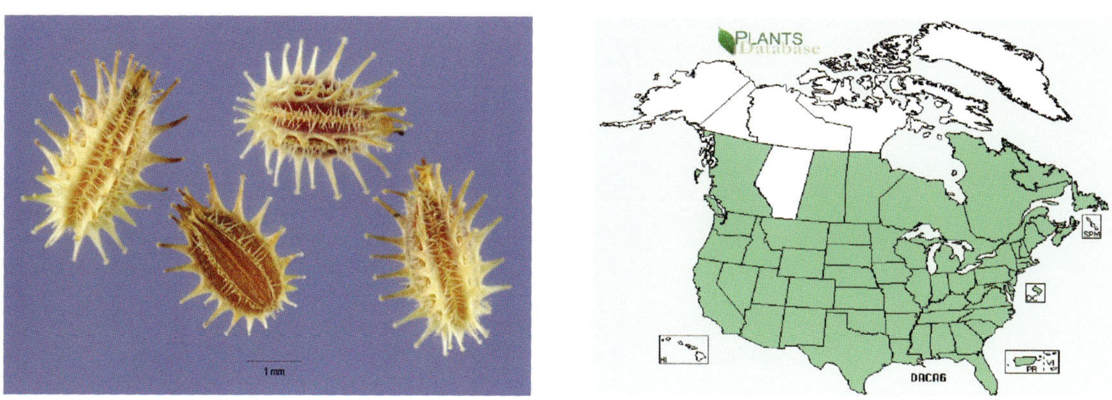

Fig. 7.48. Wild Carrot (Carrot Family) [*Daucus carota* (L)]. Other common names: Queen Anne's Lace, Bird's Nest. Oval in outline, plano-convex in cross section. Convex side bears four rows of spines. Grayish yellow to yellow with red markings. Length: 2.5–3.4 mm. Width: 1.3–2.0 mm. Introduced from Europe. Map and image courtesy of USDA-NRCS.

Fig. 7.49. Wild Garlic (Lily Family) [*Allium canadense* (L)]. Other common names: Meadow Garlic, Wild Onion. Bulblets roughly oval in outline, circular in cross section. Yellowish to light brown, with a shiny, lightly scored surface. Length: 5.0–10.0 mm. Width: 3.0–6.0 mm. The bulblet has a pungent garlic odor when crushed. Native to Eurasia. Map and image courtesy of USDA-NRCS.

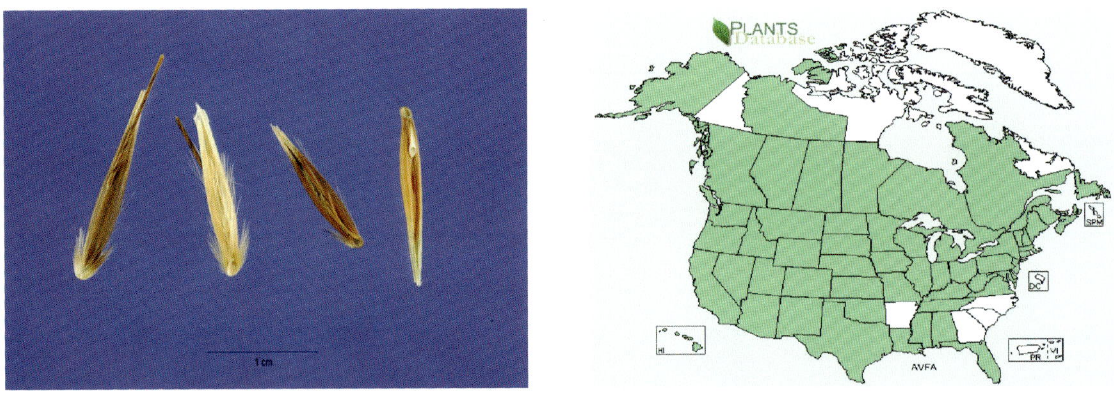

Fig. 7.50. Wild Oat (Grass Family) [*Avena fatua* (L)]. Other common names: Flaxgrass, Oatgrass, Wheat Oats. Narrowly elliptical in outline, kidney-shaped in cross section. May have a long, dark, twisted awn; hairs may be present. Grey, brown, or yellow. Length: 11.0–14.0 mm. Width: 2.0–4.0 mm. Naturalized from Europe. Map and image courtesy of USDA-NRCS.

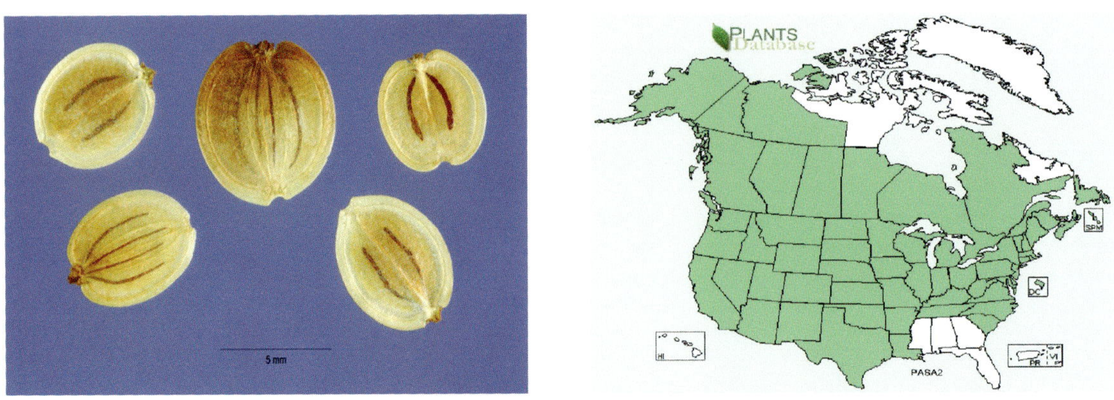

Fig. 7.51. Wild Parsnip (Carrot Family) [*Pastinaca sativa* (L)]. Other common names: Field Parsnip, Parsnip. Oval, almost circular in outline. Flat in cross section. Golden, with dark brown stripes on each side. Margins winged. Length: 5.0–7.0 mm. Width: 3.7–5.5 mm. Native to Europe. Map and image courtesy of USDA-NRCS.

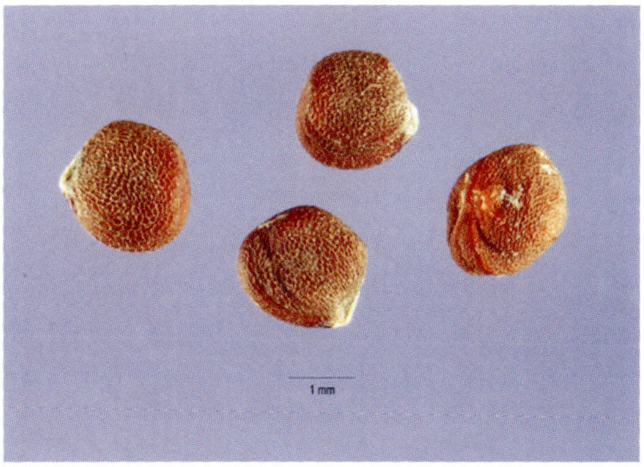

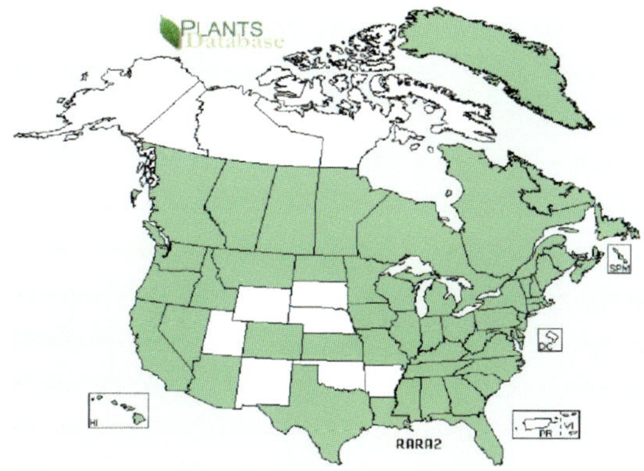

Fig. 7.52. Wild Radish (Mustard Family) (*Raphanus raphanistrum*). Other common name: none. Pod long, ribbed, and pinched between the seeds. Seed oval, with a groove along one side. Surface is granular. Golden brown to brown. Length: 2.0–3.8 mm. Width: 1.6–2.2 mm. Seed may be toxic. Map and image courtesy of USDA-NRCS.

8 Minerals of Agricultural Importance

Calcium

Numerous ingredients are used to add calcium to animal rations. Three of the most common are calcium carbonate, calcium chloride, and calcium sulfate.

Definitions

Calcium carbonate is a product true to name that contains a minimum of 38% calcium.

Calcium carbonate ($CaCO_3$) is taken by open quarry and underground mining. The raw material is crushed, dried, and screened to the desired particle size. Calcium carbonate deposits are located across the United States.

Calcium chloride is the calcium salt of hydrochloric acid, generally expressed as $CaCl_2$ and its hydrated forms.

Calcium chloride is water soluble and used in liquid feeds as a water-soluble source of calcium. There are three different processes used in the production of calcium chloride. One process produces calcium chloride as a co-product with soda ash. Another process reacts calcium carbonate with hydrochloric acid to produce $CaCl_2$. A natural brine process is also used in the production of $CaCl_2$.

Calcium sulfate is the calcium salt of sulfuric acid, generally expressed as $CaSO_4$ and its hydrated forms.

Calcium sulfate is used as a source of calcium and sulfur. Calcium sulfate is a naturally occurring material found across the United States. It is mined, crushed, dried, and screened to the desired particle size.

Macroscopic Features

Calcium carbonate can vary from white to gray in color and from a very fine powder to granular in particle size. For use in the feed industry, coarse particles and fine material should not be in the same sample. This should be avoided in order to reduce segregation in the food.

Calcium chloride is white and usually comes as a flake or powder. It is very hygroscopic and will form lumps and chunks if exposed to the air very long.

Calcium sulfate will vary from a white powdery crystal to a grayish particle depending on the source. The particle size and chemical make-up can also vary significantly depending on geographic location, even within the same company.

Stereoscopic Features

Calcium carbonate (Figs. 8.1–8.3) can vary greatly under the microscope. Depending on its source, it is generally an opaque, granular particle. Some powders are used, but usually a 20- to 200-mesh product is used in the feed industry. The particles are irregular in shape.

Calcium chloride will come as a powder or flake. If either product is exposed to the air for any period of time, it will start to look "wet" due to the hygroscopic nature of calcium chloride. The flake particles are fairly large (>20 mesh) and are irregularly shaped. The powder contains small white granular particles.

Calcium sulfate particles (Fig. 8.4) vary from a fine powder to a small grainy particle.

Fig. 8.1. CCC Lime.

Fig. 8.2. Unical S.

Note: Red arrows represent a 1-mm length.

Fig. 8.3. Con-Lime.

Fig. 8.4. $CaSO_4$.

Many times this variation will occur within the same sample of material. When observed through the microscope, the fines can be seen coating the larger particles. Specks of gray are obvious in many of the sources. This is due to some of the impurities that may be present when this material is processed.

Cobalt

There are two main sources of cobalt in the feed industry: cobalt carbonate and cobalt sulfate.

Two different hydrated forms of cobalt sulfate are available: cobalt sulfate monohydrate and cobalt sulfate heptahydrate. The monohydrate form is used almost exclusively.

Definitions

Cobalt carbonate is the cobalt salt of carbonic acid, generally expressed as $CoCO_3$ and its hydrated forms.

Cobalt carbonate is made synthetically by dissolving cobalt metal in sulfuric acid. The $CoSO_4$ is precipitated, filtered, dried, and pulverized before it is packaged.

Cobalt sulfate is the cobalt salt of sulfuric acid, generally expressed as $CoSO_4$ and its hydrated forms.

Cobalt sulfate is made synthetically by dissolving cobalt metal in sulfuric acid. The $CoSO_4$ is precipitated, filtered, dried, and ground. The only difference between cobalt sulfate monohydrate and cobalt sulfate heptahydrate in the production process is the amount of drying on the precipitate.

Macroscopic Features

Cobalt sulfate monohydrate and cobalt carbonate are both pink-to-lavender fine powders. Cobalt sulfate heptahydrate is a pink crystal. Most sources of cobalt ingredients are of excellent quality. If there is any sign of foreign material or contamination, it should be perceived as a problem and investigated further.

Stereoscopic Features

Cobalt carbonate crystals (Fig. 8.5) are more lavender in color; cobalt sulfate monohydrate crystals (Fig. 8.6) are more pink. $CoCO_3$

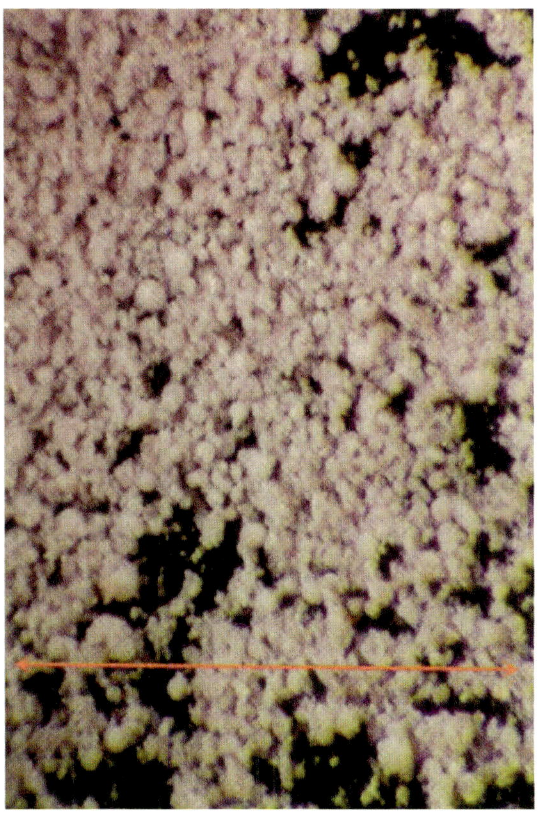

Fig. 8.5. $CoCO_3$.

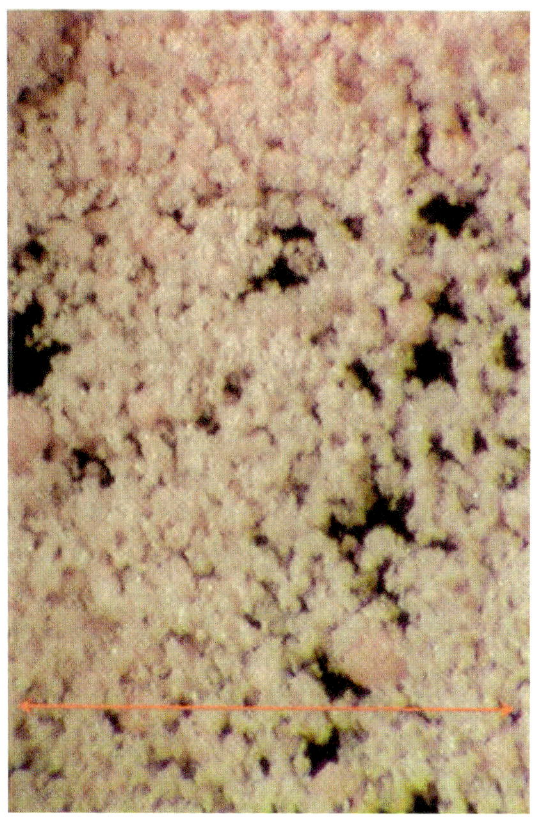

Fig. 8.6. $CoSO_4$.

crystals are slightly larger than $CoCO_3$-H_2O crystals. They are very hard to distinguish from each other unless both samples are compared to each other side by side.

If a cobalt product is examined and does not have a pure pink appearance it could be cobalt oxide or spent cobalt catalyst, which is unacceptable to the feed industry. These types of ingredients, most likely, will have black, brown, or some other dark color associated with them.

Copper

There are two main sources of copper in the feed industry: copper oxide and copper sulfate. There are a number of by-product copper sources available and as a result economically favorable sources of copper must be scrutinized for contaminants and quality issues.

Definitions

Copper oxide is the oxide form of copper, generally expressed as CuO or Cu_2O.

Copper oxide is usually produced by roasting copper metal or copper cement in an oxidizing furnace. Copper cement is the product of smelting secondary copper by-products or copper-bearing ore.

For many years, cupric oxide (CuO) was used almost exclusively in the feed industry. Recent research indicates that cuprous oxide (Cu_2O) may have some benefit. Inconsistent data from research trials has compelled many to use copper sulfate except where a water-insoluble copper source is needed.

Copper sulfate is the copper salt of sulfuric acid, generally expressed as $CuSO_4$ and its hydrated forms.

Copper sulfate pentahydrate is the most common form of $CuSO_4$ used in the feed industry. Copper sulfate is produced by dissolving copper metal or copper cement in sulfuric acid and precipitating the $CuSO_4$ crystals out of the solution. The crystals are dried and screened in order to get the correct particle size for feeding. Nearly 100% should pass through a 20-mesh screen.

Macroscopic Features

Cupric oxide should appear as a uniform black powder, cuprous oxide as a uniform reddish brown powder. Samples that contain oversized particles or obvious contamination should go through a complete chemical analysis to evaluate the quality of the product.

Copper sulfate is the most recognizable of all trace mineral ingredients. The blue, geometrically shaped crystals are easy to see in premixes and even finished feeds. There should be a specification on the amount of oversized particles (>20 mesh) to prevent potential toxicity in poultry, as well as the fines (<100 mesh) to reduce the amount of dusting when handled by employees. From a handling point of view, the $CuSO_4$ should be free flowing and not show any tendency to cake or set up.

Stereoscopic Features

Under the microscope, cupric oxide (Fig. 8.7) will appear as gray to black particles; cuprous oxide as reddish brown particles. Both should be consistent in size and color. Metallic copper, if present, should be at very low levels (<1%) and appear as finely divided copper particles. If brass or copper colored metal flakes, spheres, or turnings are visible under

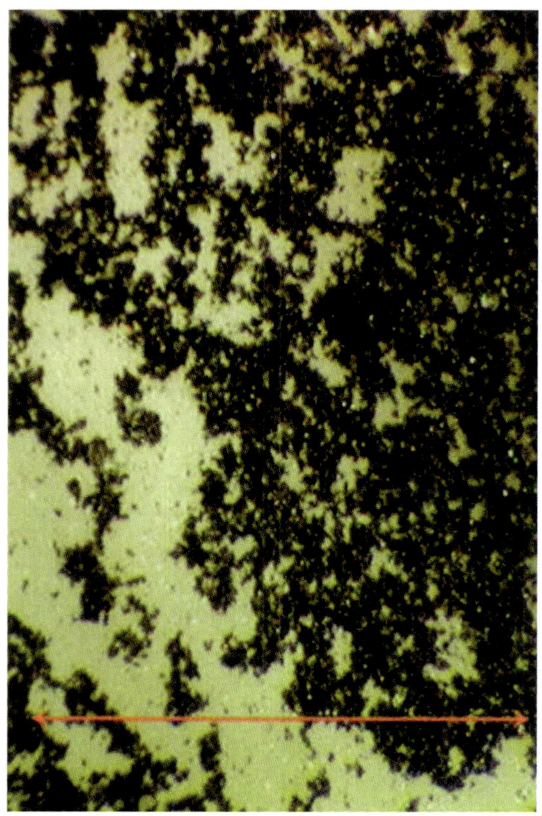

Fig. 8.7. CuO #2 Dark.

Fig. 8.8. $CuSO_4$ OB.

Fig. 8.9. $CuSO_4$ PD.

low magnification, the samples should be considered poor quality and subjected to a full chemical analysis.

Good quality copper sulfate (Figs. 8.8, 8.9) appears as translucent, light blue crystals. Consistent crystal size and shape to a certain extent indicates an in-control manufacturing process. Many producers add an anti-cake agent to prevent caking. The anti-cake agent appears as a fin, light colored powder that clings to the surface of the copper sulfate crystal (Fig. 8.10).

Copper sulfate that has a shiny glistening on the surface of the crystal indicates that the free acid was not properly cleaned from the crystal during manufacturing (Fig. 8.11). This will cause the paper bags to deteriorate and the product to set up prematurely. Copper sulfate that takes on a green cast indicates the presence of iron and other impurities (Fig. 8.12). Foreign material may be visible. White opaque crystals the same shape as the copper sulfate may be over-dried crystals (Fig. 8.12). Other foreign particles should not be present

Fig. 8.10. $CuSO_4$ PBT.

Fig. 8.11. $CuSO_4$ 97192.

Fig. 8.12. $CuSO_4$ OC97431.

(Fig. 8.13). The presence of unreacted copper metal is unacceptable. Samples that seem to have an extremely large variation in particle size (Fig. 8.14) may indicate the production process was not in control and other quality issues may be present.

Fig. 8.13. $CuSO_4$ 97312.

Iodine

There are three main sources of iodine in the feed industry: calcium iodate, potassium iodide, and ethylenediamine dihydroiodide (EDDI). Calcium iodate is the most stable. Potassium iodide is used when a water soluble product is required, and EDDI is used primarily in ruminant diets.

Definitions

Calcium iodate is the calcium salt of iodic acid, generally expressed as $Ca(IO_3)_2$ and its monohydrate.

The raw material for all three ingredients is resublimed iodine. To manufacture calcium

Fig. 8.14. $CuSO_4$ OC97312 Poor.

iodate, the resublimed iodine is dissolved in acid, oxidized to the iodate state, and reacted with calcium chloride. The resultant salt is precipitated, centrifuged, and dried.

Potassium iodide is the potassium salt of hydriodic acid, generally expressed as KI.

There are at least two different manufacturing processes used in the production of potassium iodide. One is a caustic reaction where a mixture of potassium iodide and potassium iodate is manufactured. The two compounds are separated during the crystallization step.

The second process is the green iron process. The crude iodine is reacted with iron and neutralized potash. The resultant potassium iodide solution is crystallized, dried, and screened.

EDDI is an organic compound of iodine, generally expressed as $C_2H_8N_2 \cdot 2HI$.

There are several processes used to make EDDI. Generally, an acidic form of the crude iodine is reacted with ethylenediamine to form EDDI.

Macroscopic Features

Calcium iodate is typically a white powder with no signs of foreign material. Improperly stored calcium iodate will begin to oxidize; this is indicated by a slight yellowing of the product.

Potassium iodide can either be a white powder or salt-sized crystal. This product will yellow when the oxidation process begins. Potassium iodide is very hygroscopic. The product will set up and become very hard chunks, especially in the hot and humid months of summer.

Usually EDDI is sold as a white crystal. As with all other iodine sources, EDDI will start to oxidize and yellow if it is not used on a timely basis.

Stereoscopic Features

Calcium iodate particles (Fig. 8.15) are very fine and tinted white to yellow. At very high magnification, a needle-like structure can be detected.

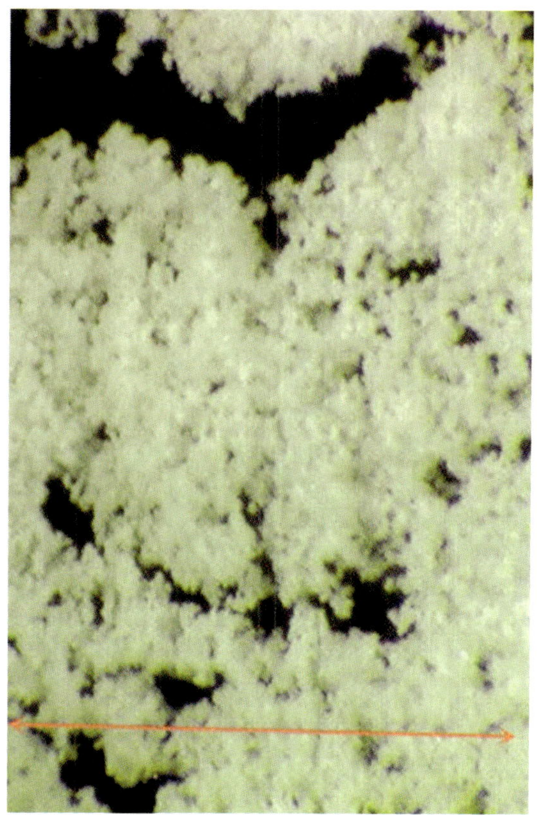

Fig. 8.15. $CaIO_3$.

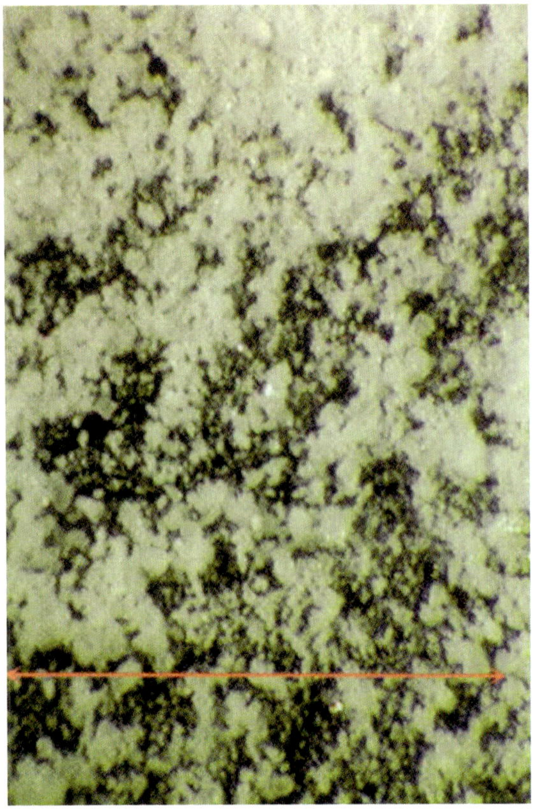

Fig. 8.16. KI.

Fig. 8.17. EDDI.

Potassium iodide crystals (Fig. 8.16) are similar in shape to salt crystals. However, KI crystals have more irregular sides and lengths to their crystal structure. KI crystals look like they have been fractured from a larger particle. They are also more opaque than salt.

EDDI crystals (Fig. 8.17) are similar in size to salt, but their corners are rounded. They are semitransparent and uniform in size and shape.

Iron

There are two main nutritional sources of iron in the feed industry: ferrous sulfate monohydrate and ferrous carbonate. Ferrous sulfate heptahydrate is used in isolated cases when a readily water-soluble source of iron is required. The feed industry also uses various iron oxides for pigmenting purposes. Although iron oxide is >50% Fe, it is not biologically available and should not be used as a nutritional source of iron.

Definitions

Ferrous sulfate is the iron salt of sulfuric acid, generally expressed as $FeSO_4$, and its hydrated forms.

As previously mentioned, the feed industry uses ferrous sulfate monohydrate and only a small amount of ferrous sulfate heptahydrate. The raw material is the same for both products. Illmenite ($FeO \cdot TiO_2$) is mined and digested in sulfuric acid to liberate the TiO_2. The g-product, $FeSO_4$, is precipitated to a wet cake which is $FeSO_4 \cdot XH_2O$. The X means the number of water molecules is very high. The wet cake is dried in a direct fired rotary drier to $FeSO_4 \cdot 7H_2O$ (heptahydrate) or $FeSO_4 \cdot H_2O$ (monohydrate).

Ferrous carbonate is the iron salt of carbonic acid, generally expressed as $FeCO_3$.

Ferrous carbonate, a naturally occurring ore called siderite, is strip mined and allowed to air dry. The raw material is screened to remove the impurities that are still attached. Finally, the raw material is ground and packaged.

Iron oxide is the oxide form of iron occurring both naturally and synthetically in various chemical valence compositions and colors, sometimes expressed as Fe_2O_3.

Both natural and synthetic iron oxides are used in the feed industry. Because they are not a nutritional source of iron, the manufacturing process is not described.

Macroscopic Features

Ferrous sulfate monohydrate is a gray granular irregularly shaped particle. Generally, the feed industry prefers as few fines (<100 mesh) as possible. This is because of the very irritating dust that can be generated while handling the product. If there are >25% fines, this dust can become a problem.

Ferrous sulfate heptahydrate is a light lime green wedge-shaped particle. With this product, caking can be a problem. The particles should be free flowing without showing any tendency to stick to one another.

Ferrous carbonate is a beige powder. Any sign of a dark brown or red color will indicate that the iron has oxidized to the ferric state and is forming iron oxide.

Iron oxides can be one of many colors. The most common color used is red. Iron oxides used for pigmenting purposes are Fe_2O_3 and are nonmagnetic. The presence of magnetic iron indicates that magnetite (Fe_3O_4) is in the sample. Both forms of iron are biologically unavailable and are not used as a nutritional source of iron.

Stereoscopic Features

Ferrous sulfate monohydrate (Fig. 8.18) appears as fairly large gray particles with smaller gray particles interspersed between the larger particles. It is possible to see very small red specks on the surface of the particles of $FeSO_4 \cdot H_2O$. This is a result of the drying process. The specks are where the iron has begun to oxidize to iron oxide. These specks are common and will not be visible with the naked eye. The presence of the specks at this level will not significantly affect quality. If the red color is visible with the naked eye, then total iron and biological available iron should be run on the sample.

Fig. 8.18. $FeSO_4$ Mono.

Ferrous sulfate heptahydrate crystals (Figs. 8.19, 8.20) are larger, semitransparent, greenish colored particles. Many of the crystals are wedge shaped. As more and more of the water is removed, the wedge-shaped crystals fracture and become crumbly like the monohydrate crystals. The red specks should not be present on the heptahydrate crystals.

Ferrous carbonate (Fig. 8.21) will appear as a tannish-brown to orange material. If the iron has started to oxidize any, the color will be dark red from iron oxide. Any amount of red iron oxide in the sample is undesirable since it is not available to the animal. Since ferrous carbonate is mined and processed as described earlier, there are periods during the year that make it difficult to remove all the impurities. The impurities will appear as foreign material in the $FeCO_3$. It is not a detrimental contamination, but the iron content should be checked by chemical analysis to make sure it meets the guarantee.

Fig. 8.19. $FeSO_4H$ without AC.

Fig. 8.20. $FeSO_4H$ with AC.

Fig. 8.21. $FeCO_3$.

Iron oxide particles are very fine. As mentioned, several different colors are used. Red iron oxide is the most common.

Red iron oxide particles will be from very dark, dull red to bright red, depending on whether it is naturally occurring or synthetic. Generally, there should not be any magnetic particles in any source of iron.

Magnesium

There are three primary sources of magnesium used in the feed industry. They are magnesium oxide, magnesium sulfate heptahydrate, and magnesium sulfate anhydrous.

Definitions

Magnesium oxide is the oxide of magnesium, generally expressed as MgO.

There are many different sources of magnesium oxide. There are two common types of production processes. The first process uses magnesite ($MgCO_3$) as the raw material. The magnesite is calcined at elevated temperatures to make magnesium oxide.

The second process uses dolomite as the raw material. Dolomite is a combination of magnesium carbonate and calcium carbonate. The dolomite is calcined to form magnesium oxide.

Magnesium sulfate heptahydrate and ***magnesium sulfate anhydrous*** are the magnesium salts of sulfuric acid, generally expressed as $MgSO_4$ and its hydrated forms.

Magnesium sulfate heptahydrate is produced by reacting magnesium hydroxide or magnesium oxide with sulfuric acid. Magnesium sulfate heptahydrate crystals are precipitated out and screened.

Magnesium sulfate anhydrous is produced from a naturally occurring ore called kiesenite. Kiesenite is a magnesium sulfate monohydrate. It is further processed and dried to form magnesium sulfate anhydrous.

Macroscopic Features

Magnesium oxide comes in many different colors and particle sizes. One of the two most common types of MgO is a beige product that ranges in particle size from 16 mesh to a fine powder. Some of the larger particles in this product can range from white to brown in color. This product typically runs 54–56% Mg on a loss free basis.

The second common source of magnesium oxide is pure white in color. It is available in many different particle size grades from a coarse prill to a <325 mesh powder.

Magnesium sulfate anhydrous (also know as Epsom salts) is a rectangular shaped semitransparent crystal. Crystal size ranges from 10 to 60 mesh. This product is hygroscropic and will set up when exposed to the atmosphere.

Magnesium sulfate anhydrous is a white chalky-looking granule. Particle size varies from 30 to 100 mesh.

Stereoscopic Features

Magnesium oxide—The beige product (Figs. 8.22, 8.23) has a wide range of particle sizes and colors under the stereoscope. It is especially evident when the sample is screened.

Fig. 8.22. MgO EZ.

Fig. 8.23. MgO Prem.

Fig. 8.24. MgO.

Fig. 8.25. $MgSO_4$ Heptahydrate.

This variation is a result of the raw material used and the calcining process. There are various levels of iron and other trace elements present in the material and this shows up as different colors between white and brown.

The white magnesium oxide (Fig. 8.24) is very pure appearing. Depending on the granulation, particle size can be a coarse (20 mesh) round prill or very fine (<325 mesh) powder. The particle size and shape is very consistent within each product. For example, there are no fines in the prilled products, nor any coarse particles in the fine products.

Magnesium sulfate heptahydrate (Fig. 8.25)—The crystals of this ingredient appear very large even at low magnification. Most of the crystals are rectangular in shape with wedge-shaped pointed ends. The crystals are nearly fully transparent when magnified. The surface is smooth for most crystals, but some crystals are slightly textured on the surface. The smallest particles present are still larger than 60 mesh.

Magnesium sulfate anhydrous (Fig. 8.26)—These crystals are white, opaque, and irregularly shaped. There are no super fine particles present. The smallest is about 100 mesh. Occasionally, a transparent crystal will be present. This would be a hydrated form of magnesium sulfate.

Mangenese

There are two main sources of manganese in the feed industry: manganous oxide and manganese sulfate. The raw material for both of these products is naturally occurring manganese dioxide ore. Since there are not any domestic sources of manganese dioxide in the United States, all of the raw material must be imported.

Definitions

Manganous oxide is an oxide form of manganese, generally expressed as MnO.

Fig. 8.26. $MgSO_4$ Anhydrous.

Manganous oxide is manufactured from manganese dioxide (MnO_2) by one of two processes. One is a batch process that uses natural gas as the reducing agent. The other process uses a carbon source such as coal as the reducing agent. The MnO_2 is mixed with the carbon source and is processed continuously through a kiln at temperatures near 1100°C. The MnO from this process must be cooled as fast as possible to prevent it from reoxidizing to MnO_2. Manganese dioxide is undesirable because it is not biologically available to the animal.

Manganese sulfate is the manganese salt of sulfuric acid, generally expressed as $MnSO_4$ and its hydrated forms.

The feed industry uses the monohydrate form of $MnSO_4$. Manganese sulfate is produced by dissolving manganous oxide in sulfuric acid. The manganese sulfate is precipitated out, dried, and screened or ground, depending on the desired particle size.

Macroscopic Features

Feed-grade MnO should be 60% Mn (minimum) with a maximum of 3% manganese dioxide. The color of manganous oxide can be an indicator of quality. Good-quality MnO has a color of olive green. If the color is more of a brown or even blackish color, it is an indication that the manganese dioxide content is too high. As with other oxide trace minerals, the absence of oversize particles (>20 mesh) is required. Preferably, most of the product will pass through a 100 mesh.

The primary characteristic to check in manganese sulfate is particle size. If a powder is ordered, no large particles should be present. Conversely, a minimal amount of fines (<100 mesh) should be present in a granular product. The color of manganese sulfate ranges from white to grayish-white depending on the source.

Stereoscopic Features

Manganous oxide produced from the batch process will be greenish in color with no presence of foreign material. Manganous oxide (Fig. 8.27) from the continuous process will have gray-to-black specks throughout. This is either the unburned carbon source or ash residue from the carbon source, which is used as the reducing agent. The main thing to watch for in manganous oxide is an overabundant amount (>3%) of manganese dioxide, which is the starting raw material. Manganese dioxide will be the same particle size as manganous oxide, but it will be a brown-to-black particle as opposed to olive green.

The manganese sulfate (Figs. 8.28, 8.29, 8.30) that enters the feed industry has been of good quality. The particles have no definite shape. They range from larger spherical particles to smaller fractured particles. The color of the particles is white to grayish-white.

Phosphorous

There are numerous ingredients used to add phosphorus to animal rations. Some of the more common ingredients used are

Fig. 8.27. MnO.

Fig. 8.28. $MnSO_4$ 32% 17×.

Fig. 8.29. $MnSO_4$ 29.5% 17×.

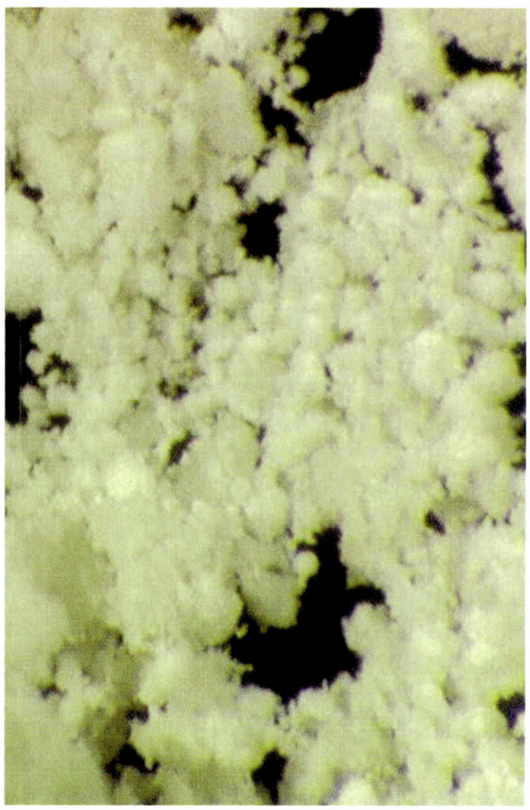

Fig. 8.30. $MnSO_4$ 90×.

defluorinated phosphate, dicalcium/monocalcium phosphate, sodium tripolyphosphate, monosodium phosphate, and monoammonium phosphate.

Definitions

Defluorinated phosphate includes either calcined, fused, precipitated, or reacted calcium phosphate. It must contain not more than one part of fluorine to 100 parts of phosphorus.

The raw material for this product is the solid material screened off during the production of phosphoric acid. This solid phosphate rock is heated in the presence of soda ash to produce defluorinated phosphate.

Dicalcium/monocalcium phosphate is a calcium salt of phosphoric acid, generally expressed as $CaHPO_4$ and $CaH_4(PO_4)_2$ and their hydrated forms.

Dicalcium and monocalcium phosphate are produced by creating phosphoric acid with limestone. These products are two of the most common sources of phosphorus. They are produced all over the United States.

Sodium tripolyphosphate is the sodium salt of phosphoric acid, generally expressed as $Na_5P_3O_{10}$.

This product is produced by reacting soda ash with phosphoric acid. Sodium tripolyphosphate is used when little or no calcium is desired in the ration.

Monosodium phosphate is the sodium salt of phosphoric acid, generally expressed as NaH_2PO_4 and its hydrated forms.

This product is produced by reacting soda ash with phosphoric acid. Monosodium phosphate is used when little or no calcium is desired in the ration. Typically, this is used more in dairy and beef cow rations.

Monoammonium phosphate is the product resulting from the neutralization of phosphoric acid, which contains not <9% nitrogen and 23% phosphorus.

This product is produced by reacting phosphoric acid with anhydrous ammonia or ammonium hydroxide. Monoammonium phosphate is used as a source of phosphorous and as a source of nonprotein nitrogen (NPN) for ruminants.

Macroscopic Features

Defluorinated phosphate has a wide range in particle size as well as color. The fines are a more brownish color while the coarse particles range from tan to brown to dark gray.

Dicalcium/monocalcium phosphate is usually a consistent looking gray granule. The particle size is typically 30 to 100 mesh. Sometimes a slightly larger granule may be present, which is a darker gray than the typical color.

Sodium tripolyphosphate is a pure white powder. It may contain a few larger particles, but they are very friable and break apart easily.

Monosodium phosphate is a pure white granular product. Most of the product will be between 20 and 100 mesh.

Monoammonium phosphate is highly water soluble. It is a coarser, mostly gray product. The top size on this product is near 14 mesh. The product is dark gray with the exception of a few tannish-brown particles interspersed throughout.

Stereoscopic Features

Defluorinated phosphate—At lower magnifications, a wide range of color and particle sizes is very evident. The larger particles are irregularly shaped and vary from almost white in color to dark gray. At higher magnifications, the crystal structures becomes very evident. The surface of the particles, especially the smaller brown particles, appears to be made up of very small, glassy spheres. It is not as evident on the coarser particles.

Dicalcium/monocalcium phosphate (Fig. 8.31)—The particles are fairly consistent in color and size. Occasionally, larger and darker particles are present. Particle shape appears to be very irregular, especially on the surface under higher magnifications.

Sodium tripolyphosphate (Fig. 8.32)—Under the stereoscope, the particle size appears to range from a very fine powder to about a 40-mesh particle. It is a pure white crystal. Under higher magnifications the finer particles appear to cling to the larger ones.

Monosodium phosphate (Fig. 8.33)—This material appears as very large, pure white, irregularly shaped crystals. There is not a large

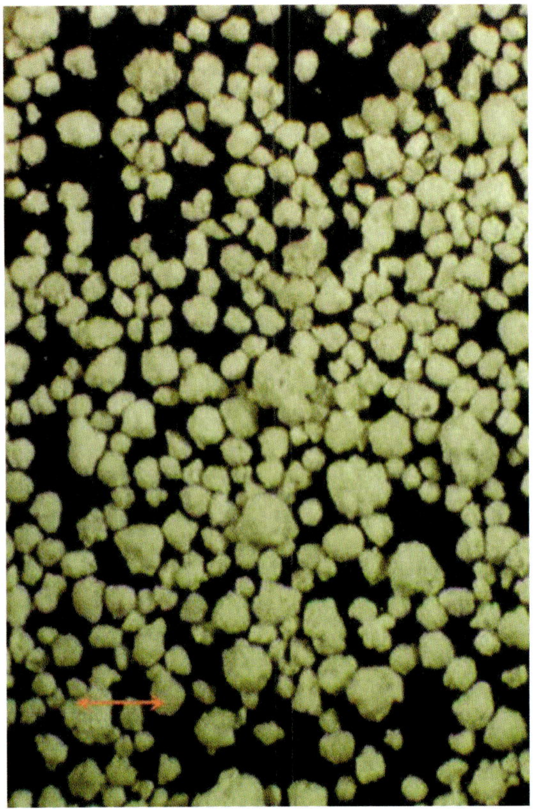

Fig. 8.31. DiCal.

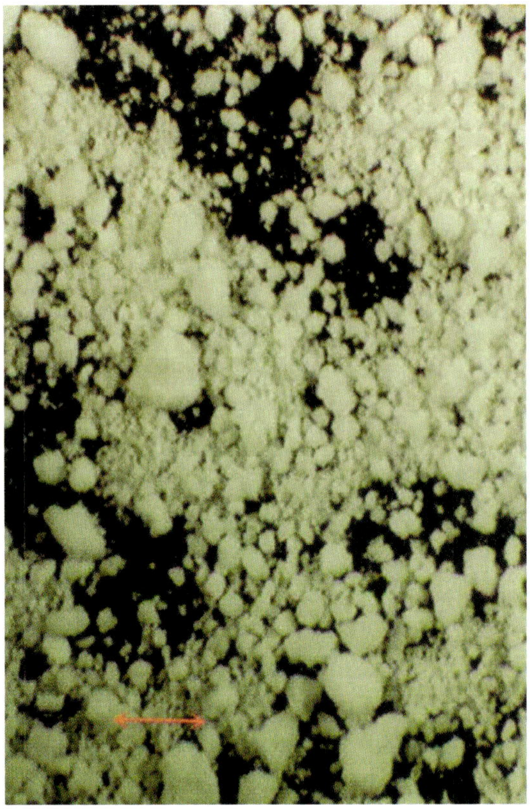

Fig. 8.32. Risodium PO_4.

presence of very fine material in this product. If there are any fines, they do not tend to cling to the larger crystals.

Monoammonium phosphate (Fig. 8.34)—These particles can be 20 mesh or larger. There are fines present in this product. The crystals are dark gray and very irregular in shape. Some crystals look fractured under the stereoscope.

Potassium

There are three main ingredients used to add potassium to animal rations. They are potassium chloride, potassium sulfate, and potassium/magnesium sulfate.

Definitions

Potassium chloride is the potassium salt of hydrochloric acid, generally expressed as KCl.

Potassium chloride exists naturally in deposits at various places in North America. One primary source mines the raw material and crushes it. A significant level of salt (sodium

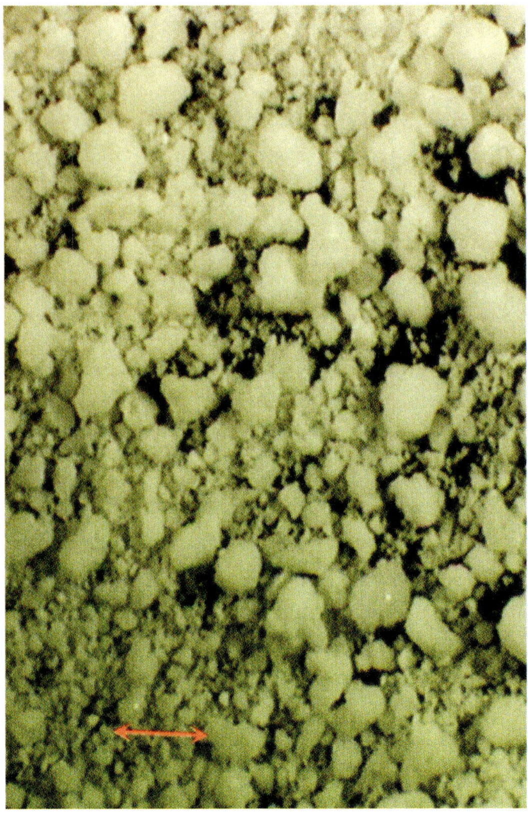

Fig. 8.33. NaH_2PO_4.

Fig. 8.34. Monoammonium PO_4.

chloride) is present and is removed through a flotation process.

A second type of process uses solution mining to remove the raw material. Water is pumped into the deposit and dissolves the potassium chloride among other compounds. The potassium chloride is precipitated out through solution chemistry.

Potassium sulfate is the potassium salt of sulfuric acid, generally expressed as K_2SO_4.

Potassium sulfate is not found naturally. It must be manufactured by reacting potassium chloride with magnesium sulfate.

Potassium/magnesium sulfate is the potassium and magnesium salt of sulfuric acid, generally expressed as K_2SO_4 and $MgSO_4$.

This is a mined product. It exists naturally in New Mexico. The mined material is high in salt. The salt is removed through a flotation process. This ingredient is used when a source of potassium, magnesium, and sulfur is desired.

Macroscopic Features

Potassium chloride—There are two very different appearing products used in the industry. The first product is a 14- to 80-mesh crystal. It is a red and white speckled product. The red crystals contain iron, which gives the red pigmentation to the crystal.

The second product is produced through solution chemistry and has had nearly all contaminants removed from it. It is a pure white, small crystal that looks like sugar. The top size on this product is 40 mesh.

Potassium/magnesium sulfate—This product is a very light red- or pink-colored crystal. Dispersed sparingly throughout are darker orange or red crystals. There is a significant amount of powder mixed in with the larger crystals. The top size of the crystals is 40 mesh.

Stereoscopic Features

Potassium chloride—The red and white speckled product (Fig. 8.35) contains a wide range of particle sizes. This is apparent under low magnifications. The red crystals appear to be more irregular in shape, and the white to clean crystals have more straight surfaces. Some of the white crystals will exhibit the cube-shaped crystal structure of salt.

The white potassium chloride (Fig. 8.36) has a more consistent and uniform particle size. The crystals are actually semitransparent instead of white. Their shape is very similar to salt. However, instead of having square corners and smooth surfaces like salt, the white potassium chloride crystals have slightly rounded corners and more of a textured surface.

Potassium sulfate—under the stereoscope the potassium sulfate crystals do not show any consistency in shape. Some crystals are roundish, some are more cube shaped, and some are totally irregular. Crystal color is various shades of light pink.

Potassium/magnesium sulfate (Fig. 8.37)—While the potassium/magnesium sulfate crystal appears to be pink to the naked eye, under the stereoscope it appears to be more of a whitish colored, semitransparent crystal. Specks of the reddish colored crystals appear throughout the material. There is no uniformity in crystal shape.

Fig. 8.35. Dyna-K.

Fig. 8.36. White KCl.

Fig. 8.37. Dynamate.

Selenium

The main source of selenium for the feed industry is sodium selenite. In higher concentrations selenium is very toxic and should be handled with extreme caution. For this reason alone, pure sodium selenite is rarely found in the feed mill. In its place are lesser concentrated selenium premixes. The sodium selenite is blended down to a safer level of selenium. Some of the more common carriers include calcium carbonate, salt, and rice hulls.

Definition

Sodium selenite is a sodium salt of selenious acid, generally expressed as $Na_2 \cdot SeO_3$ and its hydrated forms.

Sodium selenite is manufactured from pure selenium metal.

Macroscopic Features

Sodium selenite is a fine, white powder. There should be no signs of contamination

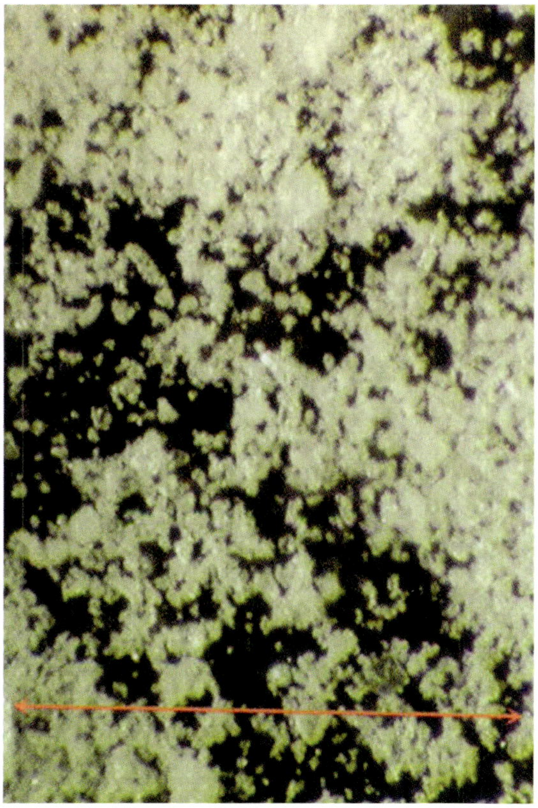

Fig. 8.38. $NaSe_2O_3$.

or foreign material. The various selenium premixes found in feed mills will have the characteristics of the carrier. Typically, the selenium content found in these premixes is 0.1% or less. Many times a red iron oxide is added to signify to plant employees that it is a selenium premix.

Stereoscopic Features

Sodium selenite (Fig. 8.38) appears as very fine, white, spherical crystals. Sodium selenite particles are hygroscopic and as a result will attach to the surface of the particles used as the carrier. When examined, the low-level selenium premixes may show evidence of red iron oxide particles.

Sulfur

There are many ingredients that carry sulfur into animal rations. Some are added specifically for the sulfur value while some carry sulfur into the mix in the form of sulfate. The more common sulfur ingredients used are granular sulfur, sulfur flour, ammonium sulfate, and sodium sulfate. There are other ingredients used and are referred to in other sections of this manual.

Definitions

Elemental sulfur, which is generally expressed as % sulfur. There are two common processes used to mine sulfur. One process pumps hot water into a mine and pushes the sulfur out. The other process recovers the sulfur from natural gas and oil fields. Both processes yield a molten liquid product. Feed grade sulfur is cooled and crushed.

Ammonium sulfate is the product resulting from the neutralization of sulfuric acid with ammonia. It contains not <21% nitrogen and not <24% sulfur. If a premix, concentrate, or supplement contains >2% of equivalent crude protein from ammonium sulfate, the label should have adequate directions for use and a prominent statement, "Caution: This feed should be used only in accordance with directions furnished on the label."

Sodium sulfate is the sodium salt of sulfuric acid generally expressed as Na_2SO_4 in its hydrated forms. Three production processes are used. One process mines the sodium sulfate from brine in Texas and California. Another process reacts salt with sulfuric acid. The third method produces sodium sulfate as a by-product in rayon and chrome industries.

Macroscopic Features

Elemental sulfur is commonly used in the form of a granule or small prill, or as a powder. Both products are yellow. The granular sulfur is a more brilliant yellow and particle size is 20–100 mesh. The powdered product is more of a dull yellow and is very fine in particle size.

Ammonium sulfate is a colorless, semitransparent crystal. The crystals are fairly coarse with most of the product falling between 14 and 40 mesh. Crystal shape to the naked eye appears cube-like to round.

Sodium sulfate is a white powder. Particle size is 60 mesh on the top size down to a fine powder.

Stereoscopic Features

Elemental sulfur—Granular sulfur (Fig. 8.39) appears as nearly perfect round spheres. The surface of the balls is smooth. Occasionally an odd-shaped particle will be present. The particles are yellow and nontransparent.

Sulfur flour appears as a very light yellow colored crystal. Under higher magnifications, the range of particle size becomes more apparent. The very fine particles appear to cling to the larger ones. The surface of the particles appears to glisten under the light source.

Ammonium sulfate (Fig. 8.40)—Particles are colorless translucent cubes with rounded corners. Fractured pieces will have irregular shapes. The top size is 14 mesh with particle as fine as 40 mesh.

Sodium sulfate (Fig. 8.41)—Particles are colorless translucent cubes with rounded corners. Fractured pieces will have irregular shapes. The top size is usually 40 mesh with particle size varying down to 100 mesh.

Fig. 8.40. Ammonium Sulfate.

Fig. 8.39. Sulfur, Prilled.

Fig. 8.41. Sodium Sulfate.

Zinc

There are two main sources of zinc in the feed industry: zinc oxide and zinc sulfate. Zinc is the most adulterated trace mineral because it makes up more of the costs of a typical trace mineral premix (TMP) and a number of by-product zinc sources are available.

Definitions

Zinc oxide is the oxide form of zinc, generally expressed as ZnO. There are a number of processes used to produce feed-grade zinc oxide. The Waelz process roasts various zinc-bearing ores in the presence of oxygen to form ZnO. The French process uses very pure zinc metal to produce a white product that is nearly 100% ZnO. Regardless of the process used, the goal is the complete conversion of the zinc to ZnO and removal of undesirable contaminants. Incomplete conversion may leave unreacted zinc metal that is not an AAFCO-approved source of zinc. Zinc oxides containing appreciable amount of zinc metal and may be considered adulterated.

Zinc sulfate is the zinc salt of sulfuric acid, generally expressed as $ZnSO_4$ and its hydrated forms. The feed industry uses a monohydrated form of $ZnSO_4$. The raw materials used to manufacture $ZnSO_4$ are generally considered to be of lower quality, secondary zinc sources. However, the manufacturers are capable of removing the impurities in the manufacturing process. The raw material is dissolved in sulfuric acid, treated to remove any impurities, and then precipitated out into $ZnSO_4$ crystals. A granular or powder $AnSO_4$ particle can be formed at this point.

Macroscopic Features

The color of zinc oxide is not a good indicator of quality. The color of good-quality ZnO can range from white, tan, brown, gray, or black. Pure ZnO is white, so any impurities present will pigment the ZnO. Impurities such as iron and manganese can pigment the ZnO very easily.

One aspect that can be checked is the absence of any large particles (>20 mesh). Usually a particle size of 100 mesh or smaller is preferred. Anything larger than this can adversely affect the biological availability of the product.

The only aspect to check in zinc sulfate is particle size. If a powder is ordered, no large particles should be present. Conversely, a minimal amount of fines (<100 mesh) should be present in a granular product. Both products are pure white-colored particles.

Stereoscopic Features

Good quality zinc oxides may vary greatly in appearance due to the different processes used to make them. The Waelz process (Figs. 8.42, 8.43) is used to make much of the product found in the feed industry. This material is primarily brownish-gray particles with black specks (coal residue) and clear glassy specks (silica).

In evaluating zinc oxide, it is necessary to look for the contaminants that indicate improper or incomplete processing. With increased magnification, zinc metal becomes

Fig. 8.42. ZnO 8000.

Fig. 8.43. ZnO 8074.

Fig. 8.44. ZnO 93345 17×.

more easily detected (Figs. 8.44, 8.45, 8.46). Separating the sample by collecting and washing the material retained on a 100-mesh sieve will more easily reveal unreacted few materials such as zinc metal, brass, or iron metal (Figs. 8.47, 8.48, 8.49). The fraction that passes 100 mesh can then be washed with water or dilute hydrochloric acid to reveal these same contaminants (Figs. 8.50, 8.51).

Unfortunately, some of the most common quality issues with zinc oxide cannot be detected under the microscope. Toxic elements such as lead, cadmium, arsenic, and mercury may be present in a sample that otherwise appears to be of good quality. Price and supplier's reputation should be considered when determining the extent of the evaluation to be performed.

Zinc sulfate appears as white to off-white hollow or solid balls under the microscope. The quality of most sources is good. $ZnSO_4$ should be nearly 100% water soluble. If it is not, this should be an indication that the material is not all $ZnSO_4$. Lead and cadmium should be monitored by chemical analysis. (Figs. 8.52–8.55).

Fig. 8.45. ZnO 93345 45×.

Fig. 8.46. ZnO 93345 90×.

Fig. 8.47. Zn Metal #016.

Fig. 8.48. Zn Metal #443.

Fig. 8.49. Zn Metal #588.

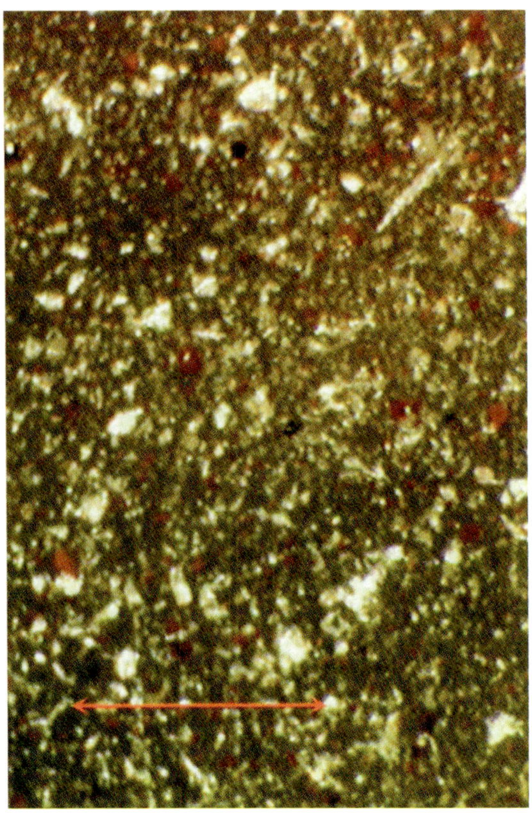

Fig. 8.50. Zn Metal #878.

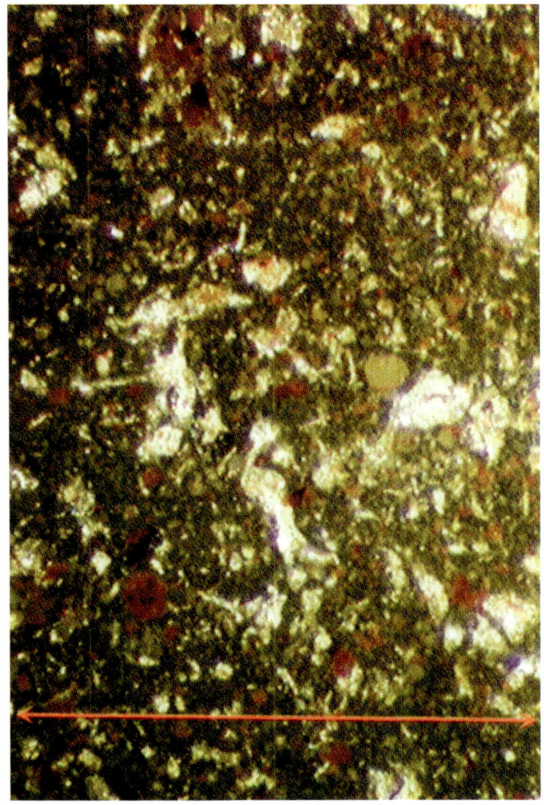

Fig. 8.51. Zn Metal #AE2.

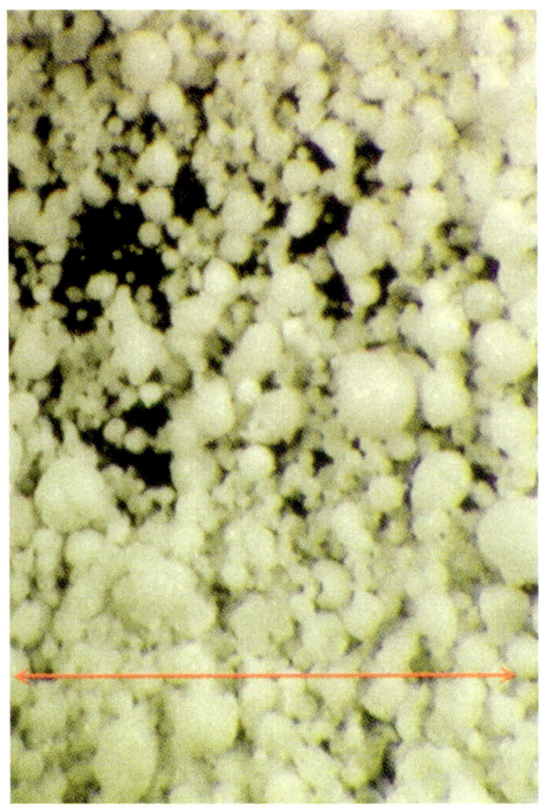

Fig. 8.52. $ZnSI_4$ 8361.

Fig. 8.53. $ZnSI_4$ 8360.

Fig. 8.54. $ZnSO_4$ 8371.

Fig. 8.55. $ZnSO_4$ 8372.

9 Microchemical Spot Tests

Microchemical Spot Tests

Organic and inorganic chemical compounds in feeds can quickly and economically be identified by microchemical spot tests. Although these are essentially qualitative, semiquantitation is possible by careful monitoring of the volume of reagents and amount of sample used, as well as previous experience.

The methodology outline is designed to help microscopists screen for medications, vitamins, preservatives, minerals, and possible contaminants by simple microchemical tests. Commercial test kits such as Quantab© (salt); Clinistick® (urine analysis test strips for sugar); and card tests for tylosin, sulfamethazine, etc. are available but can be very expensive.

Spot testing involves reacting a small sample with a reagent to form a distinct color complex or precipitate which can be observed with the naked eye, under a stereomicroscope and/or compound microscope. When only one or two crystals of analyte are available, the reaction is best performed on a glass slide under a stereomicroscope or compound microscope.

These tests may be done on spot plates (glass or porcelain), filter paper, watch glasses, petri dishes, slides, or small beakers or test tubes. Forceps, small spatulas, and dropper bottles with rubber bulb pipettes for storing reagents are required tools.

Special Analytical Methods Warning

The solvents and chemicals used in the operations and tests described in this manual have been used in analytical chemistry laboratories for many years. The degree of toxicity of chemicals ranges widely, from minor to extremely toxic. To avoid injury or health problems resulting from improper exposure to these materials, it is essential that the appropriate regulations be followed.

Before performing any chemical tests, consult the Material Safety Data Sheets (MSDS), or the equivalent, to obtain the proper handling and storage conditions of the chemicals to avoid injury or health problems. The MSDS can be obtained from the manufacturer or retail agencies.

General Procedures for Spot Tests

All twelve wells of a spot plate are filled with a test reagent, and then a small amount of finely ground sample (particle size of less than 0.75 mm) is sprinkled onto the reagent surface. The chemical reaction is observed through the microscope; resulting spots or precipitates are compared with the information in Tables 9.A, 9.B, 9.C, and 9.D. The colors indicated in these tables are only guidelines; it is essential

Text continued on p. 165

Table 9.A. Spot Tests for Drugs—Color Reaction of Pure Drugs and Detection Limits in Chop Feed.

Analyte	DMF & KOH Method I	Ehrlich's Reagent and Acetate Buffer Method II	Sakaguchi Reagent Method III	Silver Nitrate in Alcohol Method IV
Amprolium	Yellow to Orange >0.05%	—	—	—
Arsanilic Acid	—	Yellow to Orange	—	—
Carbadox	Orange Yellow 0.005%	—	—	—
Chlortetracycline HCl	—	—	Blue Violet 0.0005%	—
Emtryl or Dimetridazole	Pink to Blue >0.05%	—	—	—
Furazolidone	Blue 0.0005%	—	Pale Yellow	Pale Yellow
Gentian Violet	—	—	Turquoise	Purple Spots Trailing
Melengestrol Acetate	Faintly Brownish Yellow	—	Blue	—
Monensin Na or Salinomycin	—	—	—	—
Morantel Tartrate	—	—	Pink to Rose 0.005%	—
Nifursol	Brick Red 0.0002%	—	—	—
Nitrofurazone	Violet 0.00005%	Pale Yellow	Brownish Yellow	—
3-Nitro-4-Hydroxyphenyl-Arsonic Acid	Yellow 0.01%	Pale Yellow	—	—
Oleandomycin	—	—	Dark Olive 0.001%	—
Oxytetracycline HCl	Greenish Yellow >0.05%	—	Red 0.005%	—
Penicillin	—	Yellow 0.025%	—	—
Piperazine	—	—	—	—
Robendine HCl	Orange if concd, Light Yellow if dilute	—	—	—
Sulfamethazine	—	Yellow 0.005%	—	—
Sulfaquinoxaline	—	Yellow 0.005%	Yellow	—
Virginiamycin	Brownish Yellow	—	—	—
Zoalene	Bright Green	—	—	—

2N NaOH Method V	Ethylenediamine Method VI	3% Vanillin Reagent Method XIII	1N NaOH & Na-4-S Method XIX
—	Yellow	—	—
—	—	—	—
Orange Yellow 0.0005%	—	—	—
Brownish Yellow	Brownish Yellow	—	—
—	—	—	—
Pale Yellow	Violet Pink	—	—
—	—	—	—
—	—	—	—
—	—	Pink to Red (This test should be performed in a glass test tube.) 0.0005%	—
—	Brick Red	—	—
—	—	—	—
Brownish Yellow	Purple Black	—	—
Yellow Spots 0.0005%	Orange Yellow	—	—
Dark ppt	—	—	—
—	Greenish Yellow	—	—
—	—	—	—
—	—	—	Bright Light Red
—	—	—	—
—	—	—	—
—	—	—	—
—	—	—	—
—	Purple	—	—

Table 9.B. Spot Tests for Vitamins—Color Reaction of Vitamins.

Analyte	DMF & KOH Method I	Ehrlich's Reagent and Acetate Buffer Method II	Sakaguchi Reagent Method III	Silver Nitrate in Alcohol Method IV	2N NaOH Method V	Ethylenediamine Method VI
Vitamin A	—	—	Capsules turn dark violet but stay intact; if crushed turn blue then brown	—	—	—
Ascorbic Acid	—	—	—	Blue-Black ppt	—	—
Menadione	Khaki to Brown	Red	Red Brown	Brick Red	Brown ppt	Dirty Green to Yellow Brown ppt
Menadione Sodium Bisulphate	Dirty Green to Brown	—	Yellow to Brown	—	Pale Yellow	Olive Green
Riboflavin	—	Yellow	Orange tailing to Yellow	—	Orange ppt	Deep Yellow tailing out from powder
Thiamine	Yellow	—	Effervesces	—	Pale Yellow	Pale Yellow

Table 9.C. Spot Tests for Minerals—Color Reactions of Minerals.

Analyte	Aqueous Silver Nitrate Method VII	Potassium Ferrocyanide Method VIII	Nitroso-R-Salt Method IX	2N NaOH & Diphenylthiocarbazone Method X	20% NaOH & Quinalizarin Method XI	Bromine Water & Starch Paper Method XVII	H_3PO_2 & Starch Paper Method XVIII
Chloride	Milky White	—	—	—	—	—	—
Cobalt	—	Green (Turquoise)	Pink	—	—	—	—
Copper	—	Reddish Brown	Brown	—	—	—	—
Iron	—	Blue	Bluish Green	—	—	—	—
Magnesium	—	—	—	—	Blue	—	—
Manganese	—	Milky White	—	—	—	—	—
Phosphate (mono and dibasic)	Yellow Needles ppt	—	—	—	—	—	—
Nitrate	Dissolves completely; no change of color in solution	—	—	—	—	—	—
Sulphate	White needles	—	—	—	—	—	—
Zinc	—	—	—	Fuchsia	—	—	—
Iodate	—	—	—	—	—	—	Blue Purple
Iodide	—	—	—	—	—	Blue Purple	—

Table 9.D. Spot Tests for Animal Products—Reactions of Suspect Animal Product Particles.

Analyte	Aqueous Silver Nitrate Method VII	HCl Method XV	Glacial Acetic Acid & H_2O Method XVI	H_2O_2 & Tetramethyl-benzidine Method XX	Fehling's Solution Method XXI	Cystine Reagent Method XXII
Bone	Particle slowly turns brown then black	Effervesces; slow action	—	—	—	—
Limestone	—	Effervesces; rapid action	—	—	—	—
Blood	—	—	—	Blue/Green with bubble formation	—	—
Cystine (Hair/ Feathers)	—	—	—	—	—	Particle turns black-brown
Gelatin	—	—	Soft & Swollen	—	—	—
Hoof	—	—	Hard	—	—	—
Horn	—	—	Hard	—	—	—
Muscle tissue	—	—	—	—	Particle stains pinkish-violet	

Continued from p. 161

to compare the color responses of the analyte with a reference standard to prevent errors.

For some analytes, one or two tests are adequate for identification. For others, however, several reagents must be used to establish the pattern of color response which is unique or distinct enough for definitive identification.

When an analyte is present as an extremely fine solid particle in minute quantities, test sensitivity can be increased by using filter paper as reagent medium. When filter paper is impregnated with reaction, it provides a large surface area for the analyte to react with the reagent, and the chemical reaction proceeds rapidly. The color does not disappear as quickly as it would with a spot plate, so the reactions are localized and discernable. The pH of a reaction mixture often determines whether a satisfactory color or precipitate is formed, so the amount of reagent used and temperature and freshness of the reagent are important. Do not flood the filter paper with reagent.

Preparation of Reagents and Method Procedures

Method I

Analytes. Amprolium, Carbadox, Emtryl, furazolidone, nifursol, nitrofurazone, 3-nitro-4-hydroxyphenylarsonic acid, oxytetracycline, virginiamycin, menadione, thiamine, menadione sodium bisulphite, robendine hydrochloride, melengestrol acetate, zoalene.

Reagents. Dimethylformamide (DMF) and potassium hydroxide (KOH).

- **a)** Dimethylformamide—reagent grade.
- **b)** 4% KOH—dissolve 4 g of KOH in 100 mL ethanol or methanol.

Procedure.

1. Just before using, mix 1 part KOH with 10 parts DMF.

2. Sprinkle analyte onto the surface of the mixture.
3. Observe color change.

Method II

Analytes. Nitrofurazone, 3-nitro-4-hydroxy-phenylarsonic acid, penicillin, sulfamethazine, sulfaquinoxaline, menadione, riboflavin, arsanilic acid.

Reagents. Ehrlich's reagent and buffer.

a) Stock solution—dissolve 5 g of p-dimethylaminobenzaldehyde in 100 mL of 95% ethanol. (The stock solution is stable for several months).
b) Ehrlich's reagent—mix 4 mL of stock solution (a) with 1 mL of concd sulphuric acid and add 10 mL of 95% ethanol.
c) Acetate buffer (pH 4)—dissolve 50 g of sodium acetate in 100 mL glacial acetic acid.

Procedure.

1. Place a small quantity of ground feed in a spot plate well.
2. Wet sample with 1–2 drops of buffer (c).
3. Add a few drops of Ehrlich's reagent (b).
4. Observe color change in the solution.

Method III

Analytes. Chlortetracycline, furazolidone, nitrofurazone, oleandomycin, oxytetracycline, sulfaquinoxaline, vitamin A, menadione sodium bisulphate, riboflavin, Gentian violet, melengestrol acetate, morantel tartrate, thiamine.

Reagent. Sakaguchi reagent.

a) Sakaguchi reagent—dissolve 1 g boric acid in 100 mL of 80% sulphuric acid. Store this reagent at 4C.

Procedure.

1. Fill spot plate with the reagent.
2. Sprinkle analyte on the surface.
3. Observe color change under the stereomicroscope (reagent is more effective when cold).

Method IV

Analytes. Furazolidone, ascorbic acid, Gentian violet, menadione.

Reagent. Alcoholic silver nitrate solution.

a) Alcoholic silver nitrate—saturated solution of silver nitrate in 95% ethanol (approximately 3%).

Procedure.

1. Place a few drops of the solution in a well of a spot plate.
2. Sprinkle sample on the surface.
3. Observe formation of a precipitate.

Method V

Analytes. Carbadox, chlortetracycline, furazolidone, nitrofurazone, 3-nitro-4-hydroxy-phenylarsonic acid, oleandomycin, menadione, menadione sodium bisulphate, riboflavin, thiamine.

Reagent. 2N sodium hydroxide (NaOH).

a) 2N NaOH—dissolve 40 g of NaOH in 500 mL distilled water.

Procedure.

1. Sprinkle sample onto a few drops of reagent.
2. Observe color change.

Method VI

Analytes. Amprolium, chlortetracycline, furazolidone, nitrofurazone, 3-nitro-4-hydroxy-phenylarsonic acid, oxytetracycline, menadione, menadione sodium bisulphate, riboflavin, morantel tartrate, thiamine, zoalene.

Reagent. Ethylenediamine, reagent grade.

Procedure.

1. In a fume hood, place a few drops of the reagent into a spot plate well.
2. Sprinkle a small amount of sample on the reagent.
3. Observe color change.

Method VII

Analytes. Mono- and dibasic phosphate, nitrate, sulphate, chloride, bone.

Reagent. 10% aqueous silver nitrate solution.

a) 10% aqueous silver nitrate—dissolve 10 g of silver nitrate crystals in 100 mL distilled water (store this solution in a dark bottle).

Procedure.

1. Place a few drops of the solution into a spot plate well.
2. Sprinkle sample onto the solution.
3. Observe the formation of colored precipitate.

Method VIII

Analytes. Cobalt salts, copper salts, iron, manganese.

Reagent. Potassium ferricyanide.

Procedure.

1. Dissolve approximately 0.003 g of potassium ferricyanide in 10 drops of distilled water on a watch glass.
2. Sprinkle sample onto the solution.
3. Observe characteristic colors unique to each mineral.

Method IX

Analytes. Cobalt salts, copper salts, iron.

Reagents. Sodium potassium tartrate solution and nitroso-R-salt.

a) Sodium potassium tartrate solution—weigh 100 g and make up to 500 mL volume with distilled water.

b) Nitroso-R-salt—weigh 1 g of 1-nitroso-2-hydroxynaphthalene-3,6-disulfonic acid (sodium salt) and make up to 500 mL volume with distilled water.

Procedure.

1. Moisten filter paper with a few drops of solution (a).
2. Sprinkle sample over the moist filter paper with a spatula.
3. Add a few drops of solution (b).
4. Observe distinct color spots unique to each mineral (Adding a drop of 6N HCl to the solution allows for a better reaction of oxides and carbonate salts. The addition of the 6N HCl to the solution and observations must be done in a fume hood as cyanide gas may be formed.).

Method X

Analyte. Zinc.

Reagents. 2N sodium hydroxide and diphenylthiocarbazone.

a) 2N sodium hydroxide—dissolve 40 g of NaOH in 500 mL of distilled water.

b) Diphenylthiocarbazone solution—add 10 mL diphenylthiocarbazone to 100 mL carbon tetrachloride in a beaker (in a fume hood).

Procedure.

1. Moisten filter paper with a few drops of solution (a). Work in a fume hood.
2. Sprinkle sample over the moist filter paper.
3. Add 1–2 drops of solution (b).
4. A fuchsia color indicates the presence of zinc.
5. A control should be tested at the same time to prevent false positives.

Method XI

Analyte. Magnesium.

Reagents. 20% sodium hydroxide and 0.05% alcoholic quinalizarin solution.

a) 20% aqueous NaOH.

b) 0.05% ethanolic quinalizarin solution.

Procedure.

1. Dissolve sample in 2–3 drops of water on a white spot plate.
2. Add 1 drop of solution (a) and 1 drop of solution (b). Allow to react.

3. Magnesium turns the solution blue.
4. Compare blue color with the violet color of a blank solution (i.e. negative control solution comprising of only the reagents) after reaction. The blue precipitate that forms if magnesium is present can be more easily seen with the aid of a stereomicroscope. A precipitate will not be present in the blank solution.

Method XII

Analyte. Urea.

Reagents. Urease and pH test strip (or pink litmus paper).

Procedure.

1. Place 0.5 g of a finely ground sample in a 10 mL Erlenmeyer flask with a glass stopper.
2. Add approximately 3 µg of urease.
3. Moisten thoroughly with distilled water. Suspend a moist pH test strip (or pink litmus paper) above the analyte by inserting the paper between the glass stopper and the neck of the flask. Take care not to touch the lower end of the pH test strip to the side of the flask.
4. Let stand up to 30 min. The paper turns blue if urea is present. The use of both a positive and a negative control is recommended.

Method XIII

Analytes. Monensin sodium salt, salinomycin.

Reagent. 3% vanillin solution.

a) 3% vanillin—dissolve 3 g vanillin in 50 mL methanol. Add 0.5 mL sulphuric acid and dilute with methanol to 100 mL. Prepare the reagent before use.

Procedure.

1. Place approximately 5 g of sample in 100 mL beaker.
2. Add 20 mL of methanol, swirl, and let stand until solution clears.
3. Decant 10 mL of the extract into 30 mL test tube.
4. Add 10 mL of vanillin solution to the extract.
5. Place test tube in 200 mL beaker containing 150 mL H_2O, preheated to 60C.
6. Let stand for 10 min.
7. Positive monensin or salinomycin yields a pink to red color, depending on concentration. The use of both a positive and a negative control is recommended.

Method XIV

Analyte. Amprolium.

Reagents. Extractant, naphthalenediol, and potassium ferricyanide.

a) Extractant—add 10 mL methanol and 10 mL chloroform to 10 mL distilled water. Shake thoroughly just before use.
b) Naphthalenediol reagent—dissolve 50 mg 2,7-naphthalenediol in 500 mL 95% methanol. Add 150 mL distilled water and mix thoroughly.
c) Potassium ferricyanide reagent—dissolve 1.25 g $K_3Fe(CN)_6$ and 3 g NaOH in 500 mL distilled water.

Procedure.

1. Add 30 mL extractant (a) to 10 g of sample and mix thoroughly.
2. Let stand for 2 minutes.
3. Filter the extract and transfer 2–3 drops of the filtrate to wells of a white spot plate.
4. Mix 3 mL reagent (b) and 1 mL reagent (c).
5. Add 3 drops of the above mixture to each well. Stir to mix.
6. Observe a change of color.
7. A reddish brown solution indicates the presence of amprolium.
8. A blank should be run simultaneously.

NOTE: When a sample contains a large amount of alfalfa or grass, the extract can be decolorized by passing the extract through a silica gel column (pH 9) before testing.

Method XV

Analytes. Bone vs. limestone.

Reagent. Hydrochloric acid.

Procedure.

1. Place the sample on a watch glass or in a well of a black spot plate.
2. Add hydrochloric acid.
3. Watch for effervescence. Limestone will react quickly, mono- and dicalcium phosphate will react slightly and momentarily, and bone will effervesce relatively slowly, but steadily.

Method XVI

Analytes. Gelatin vs. hoof or horn.

Reagent. Acetic acid solution.

a) Mix glacial acetic acid 1:1 (v/v) with distilled water.

Procedure.

1. Cover sample with glacial acetic acid solution (a) in a spot plate.
2. Check the sample in 1 h.
3. Gelatin will be soft and swollen, while hoof and horn will remain tough.

Method XVII

Analyte. Iodide.

Reagents. Bromine water and starch paper.

a) Bromine water—dilute 1 mL of saturated bromine water to a final volume of 20 mL with distilled water.

Procedure.

1. Moisten starch paper with dilute bromine water (a).
2. Sprinkle the fine particles of the sample on the starch paper.
3. A blue-purple color indicates the presence of iodide.

Method XVIII

Analyte. Iodate.

Reagents. Hypophosphorus acid (30–32%) and starch paper.

Procedure.

1. Moisten starch paper with the acid.
2. Sprinkle the fine particles of the sample on the starch paper.
3. A blue-purple color indicates the presence of iodate. This reaction may take a few minutes.

Method XIX

Analyte. Piperazine.

Reagents. 1N NaOH and Na β-napthoquinone-4-sulfonate.

a) 1N NaOH—weigh 4 g into 100 mL distilled water/ethanol (1:1, v/v).

b) Na β-napthoquinone-4-sulfonate—dissolve 30 μg in 20 mL distilled water.

Procedure.

1. Add 4 drops of (a) followed by 4 drops of (b) onto filter paper.
2. Sprinkle sample over the wet surface of the filter paper.
3. A splash of bright light red color indicates the presence of piperazine.

Method XX

Analyte. Blood.

Reagents. 3,3′,5,5′-tetramethylbenzidine, glacial acetic acid, and hydrogen peroxide.

a) 3,3′,5,5′-tetramethylbenzidine—mix 1 g with 100 mL glacial acetic acid. Dilute with 150 mL distilled water.

b) 3% H_2O_2—dilute stock H_2O_2 solution to 3% with distilled water.

Procedure.

1. Just prior to use, mix 4 parts of (a) with 1 part (b).

2. Place a few suspect blood particles on a microscope slide.
3. Add a drop of the reagent. If blood is present, a blue color will develop around the particle.
4. If observing the reaction with a compound microscope, crush the suspect blood particles on the microscope slide and cover with a coverslip prior to adding the solution (1). A blue/green color will form around blood and the O_2 bubbles will be captured under the coverslip and visible.

Method XXI

Analyte. Muscle tissue.

Reagent. Fehling's solution (2 stock solutions).

a) Copper (II) sulphate pentahydrate solution—dissolve 69.3 g $CuSO_4 \cdot 5H_2O$ in 1 L of distilled water.
b) Potassium sodium tartrate tetrahydrate solution—dissolve 346 g $KNaC_4H_4O_6 \cdot 4H_2O$ and 120 g NaOH in 1 L of distilled water.

NOTE: The formation of small cubic red copper (I) oxide crystal occurs in Fehling's reagent that has been stored for too long. Although this does not inhibit the reaction, a fresh solution is preferred.

Procedure.

1. Just prior to use mix equals parts (1:1, v/v) of stock solution (a) and (b) to compose the Fehling's reagent.
2. Add few drops of the reagent on a microscope slide.
3. Place a few particles from the flotate on the slide.
4. Place a coverslip and observe under the compound microscope. If muscle fibers are present, they will stain pale pinkish-violet.

Method XXII

Analyte. Cystine containing structures (e.g. hairs and feathers—raw or hydrolyzed).

Reagent. Cystine reagent.

a) Dissolve 2 g of lead acetate in 10 g NaOH in 100 mL distilled water.

Procedure.

1. Place the suspect feather or hair fragment on a slide.
2. Add few drops of the reagent on a microscope slide.
3. Overlay with a coverslip. Carefully heat the slide while avoiding boiling. Cystine containing structures will turn black-brown.

NOTE: Heating is not mandatory, it just accelerates the reaction. The reaction will take about 10–20 min without heating.

Appendix A: Glossary

This section contains a glossary that defines the special nomenclature of agricultural microscopy drawn from the various scientific disciplines. Suggested reference sources and related information are also listed.

Achene—A small, dry, thin-walled fruit that does not split open when ripe.

Acuminate—Gradually tapering to a sharp point.

Acute—Sharp-pointed but less tapering than acuminate.

Adjunct—An alternative carbohydrate source in the brewing industry.

Adulterant—Any foreign or extraneous material that is added purposely to reduce quality or to make a product inferior. In addition, see AAFCO Adulterant Regulation, Section 7.

Aleurone—Protein matter occurring in the form of minute grains or granules in the seeds of numerous plants.

Amorphous—Having no definite form.

Apex—The tip or point.

As is fraction—The fraction of a sample that has not been subjected to a floatation procedure.

Awn—One of the slender bristles constituting the beard of cereal grains.

Axial—Of or pertaining to an axis.

Axil—The angle between a leaf or bract and the stem.

Basal—Located at the base.

Birefringence—The double refraction of light in a transparent, molecularly ordered material, which is manifested by the existence of orientation-dependent differences in refractive index.

Biconvex—Convex on both sides, lens-shaped.

Bract—A reduced leaf-like structure in the axil where a flower or flower cluster arises.

Calyx—The outermost cycle of floral parts, consisting of sepals.

Canaliculus (plural ***Canaliculi***)—Minute canals, especially in bone.

Carpel—One of the seed-containing divisions in a multi-seeded pod, such as the sections of an orange, which began as a division of the ovary.

Caryopsis—A fruit developed from a single carpel with pericarp united to the seed, the fruit of cereals and grasses, or a dry one-seeded fruit where the ovary wall and seed coat are fused.

Chelae—Pincers from insects or crustaceans.

Chelator—A substance that binds particular ions.

Chitin—A horny substance forming the harder part of the outer integument of insects and crustaceans.

Chromatophore—A color-producing body, such as a chloroplast or chromoplast in plant cells.

Clavate—Club-shaped.

Coarse Grains—Coarse grains are corn, sorghum and millet.

Collenchyma—Tissue made up of elongated cells with somewhat thickened walls (see ***Hypoderma***).

Conchoidal—A break that resembles a portion of a shell (e.g. glass fractures).

Contaminant—Foreign material that occurs naturally, or cannot be avoided in good factory practice.

Cortex—The outer or superficial part of a body structure.

Cortical—Of or pertaining to the cortex.

Cotyledon—Seed leaf, the first leaf or leaves developed in seed plants.

Crumb Structure—The open or closed cellular appearance produced by yeast on doughs, and fixed during baking processes.

Cuticle—The tough and flexible outer layer of an organism or part of an organism.

Cultivar—A cultivated variety of a plant distinguished from a wild type.

Dendrite—Branched outgrowths or projections of neuronal cells resembling a tree-form.

Diaphysis—The long and narrow central portion of a long bone.

Dicotyledon—A seed plant which has two seed leaves (beans, peas).

Distal—Located at the tip.

Dorsal—Designating the surface turned away from the axis, also the top side of an object.

Eliptical (***Ellipsoidal***)—In the shape of a slightly flattened circle.

Embryo—That part of the seed or grain which contains the structures that give rise to the new plant.

Endocarp—The inner layer of the pericarp.

Endosperm—The stored food in a seed which lies outside the embryo.

Epicarp—The outer layer of the pericarp.

Epidermis—The thin layer of cells forming the external integument of coat in certain structures (leaves, seeds) of seed plants.

Epiphisis (plural ***Epiphises***)—The rounded end of a long bone.

Epithelium—A cellular membrane-like tissue covering a free surface or lining a cavity, and consisting of one or more layers of cells with little intercellular substance.

Erose—Having an irregularly notched margin.

Exoskeleton—external skeleton from insects, crustaceans or mollusks.

Facet—A small, planar surface, usually one of several.

Filiform—Thread-like.

Fines—That portion of a feed sample which passes through the finest mesh sieve and is caught in the pan.

Fissure—A narrow crack or cleft.

Flotate—Material floating on the surface of a liquid (e.g., organic solvent) due to the relative density of the material to the liquid.

Floatation—A sample preparation technique that uses an organic solvent, or solvents, to separate a sample into two or more fractions based on density.

Friable—Readily crumbled or brittle.

Funicle—A small cord.

Fusiform—Spindle-shaped, tapered at both ends.

Globose (***Globular***)—Spherical.

Glume—The pair of bracts at the base of a spikelet or seed in the cereal family.

Granular—Surface roughened by minute granules.

Groat—The kernel of grain with the hull removed.

Haversian Canal—The small tube or canal through which blood vessels pass in bone.

Hilum— A scar left where the seed stalk breaks away or where the seed was attached directly to the placenta.

Histology—A study of the minute or cellular structure of animal or vegetable tissues.

Hypoderma—Any layer of tissues beneath the epidermis, serving to add strength.

Iridescent—Property of surfaces which change color as the angle of view changes as seen in soap bubbles and abalone shells.

Isodiametric—With equal diameters.

Keratin—A protein that forms the outer layer of epidermal structures such as horn, hair, and feathers.

Kernel—A whole seed of a cereal grain; dehulled seed.

Lacuna—(plural ***Lacunae***) A minute cavity in bone that is or was occupied by an osteocyte (bone cell).

Lamella—A thin plate, leaf, or layer.

Lemma—The outer bract of the flower of grasses, sometimes called a flowering glume.

Lenticular—Lens-shaped.

Loculus—One of the cells of a compound ovary.

Lumen—The opening through the center of a structure.

Macroscopic—Large enough to be viewed with the naked eye, or very slight magnification

Malleable—Capable of being molded.

Medulla—The inner core or central tissue of an organ or part.

Melanophore—Pigment cells that permit color change in animal skin or hair or scale.

Mesocarp—The middle layer of the pericarp.

Metaphysis (plural ***Metaphyses***)—the wider part of a long bone located between the epiphyses and the diaphysis.

Micron (μm)—1/1000 of a millimeter.

Micropyle—A minute opening in a seed through which the primary root emerges.

Monocotyledon—A plant which produces only one seed leaf (a grass).

Morphological—Dealing with the form and structure of animals and plants.

Mosaic—Comprising, or appearing to comprise, of diverse elements of various shapes.

Mounting medium—A liquid used to disperse and carry material on a slide for examination under the compound microscope.

Noxious weed—A plant that is difficult to control with normal cultural practices.

Nucellus—The central part of the body of an ovule, or kernel of grain containing the embryo sac.

Obovate—Egg-shaped in outline with the narrow end attached to the stalk.

Offal—Inedible (for food) parts of a butchered animal.

Osseous—Of or pertaining to bone.

Osteocyte—A branched cell embedded in the calcified matrix of bone tissue.

Otolith—A calcareous concretion in the internal ear of a vertebrate, or in the otocyst of an invertebrate.

Ovate (***Ovoid***)—Egg-shaped, the broader end at the base.

Ovule—A minute structure in seed plants, containing the embryo sac and surrounded by the nucellus, that develops into a seed after fertilization.

Palea—The tiny upper bract that combines with the lemma to enclose the kernel of a cereal.

Parenchyma—Tissue composed of thin-walled cells.

Pedicel—The stalk of a spikelet.

Pericarp—The mature and variously modified walls of the ovary.

Perisperm—Nutritive tissue of seed derived from the nucellus, and occurring outside the embryo sac.

Prion—An infectious particle composed primarily of protein.

Rachilla—The small stem on which the flower or floret is borne.

Rachis—The main stem or elongated axis of a flower arrangement or the distal portion of a feather shaft.

Radial—arranged like spokes of a wheel.

Radicle—The rudimentary root of an embryo.

Reticulate—Having a network-like appearance.

Riffler—A piece of lab equipment consisting of a series of shuts used to separate a sample into two or more representative subsamples.

Sarcomere—Basic subunit of a striated muscle tissue fiber composed of myosin and actin.

Sclerenchyma—Tissue composed of cells which have the cell wall thickened and lignified.

Scutellum—The cotyledon of a cereal grain.

Seed—A plant at rest.

Sepal—One of the members of a calyx.

Serrate—Notched or toothed on the edge.

Sclereid—A thick-walled lignified sclerenchyma cell.

Sinuous—Winding or wavy.

Sinus—A cavity or depression.

Small Grains—The group of cereals that include: wheat, barley, rye, triticale, rice, and oats.

Spermoderm—The testa or seed coat.

Spikelet—The unit of flower arrangement in cereals consisting of two glumes and one or more florets.

Stereoscope—A microscope having two optical systems which give depth to the object being viewed, or the appearance of solidity.

Stomata—Small, mouth-like openings in the epidermis of plants.

Stone Cells—A nearly isometric sclereid that is found in certain fruits.

Stratified—Arranged in layers.

Striate—Marked with slender, longitudinal grooves or lines.

Transverse—Across, horizontal.

Trichome—Any epidermal hair of a plant.

Twin Hairs—Double trichomes found on sunflower hulls.

Undulate—Wavy-surfaced.

Unicellular—One-celled.

Vacuole—A small cavity surrounded by a membrane.

Ventral—The surface of a grain that faces the stem or rachis.

Vermiform—Worm-like in shape.

Weed—A plant that is not valued where it is growing.

Appendix B: Mesh Size Conversion Table

US Mesh Size	mm	in.
3	7.08	0.279
4	4.74	0.187
6	3.36	0.132
8	2.38	0.097
10	2.00	0.075
12	1.68	0.0661
14	1.41	0.0555
16	1.19	0.0469
18	1.00	0.0394
20	0.84	0.0331
25	0.71	0.028
30	0.59	0.0232
40	0.42	0.0165
50	0.297	0.0117
60	0.25	0.0098
70	0.21	0.0083
80	0.177	0.007
100	0.149	0.0059
120	0.125	0.0049
140	0.105	0.0041
200	0.074	0.0029
250	0.062	0.0024
270	0.053	0.0021
325	0.044	0.0017
400	0.037	0.0015
450	0.033	0.0013
500	0.025	0.0010
635	0.020	0.0008

Bibliography

Allen , R. M. *Photomicrography.* D. Van Nostrand, New York, 1941.

AOAC International – *Official Methods of Analysis.* 15th Edition. AOAC International. Arlington, VA., 1990.

Association of American Feed Control Officials (AAFCO). *Official Publication of the Association of American Feed Control Officials,* 2010.

Bates, L.S., L. Barefield, W. Landgraf, R. Sample, B. Bax and S. Jaconis. *Manual of Microscopic Analysis of Feedstuffs,* 3rd ed. American Association of Feed Microscopists,1992.

Berg, G.L. ed. *Farm Chemical Handbook.* Meister Pub., Willoughby, OH, 1964.

Blanch, F.C., ed. *Handbook of Food & Agriculture.* Reinhold Pub., New York, 1955.

Canadian Food Inspection Agency (CFIA). 2007. Enhanced Health Protection from BSE. Accessed August 5, 2010 from: http://www.inspection.gc.ca/english/anima/heasan/disemala/bseesb/enhren/publie.shtml

Canadian Food Inspection Agency (CFIA). 2009. Bovine Spongiform Encephalopathy (BSE). Accessed August 5, 2010 from: http://www.inspection.gc.ca/english/anima/disemala/bseesb/bseesbfse.shtml

Chamot, E.M., Mason, C.W. *Handbook of Chemical Microscopy,* Vol. I, 3rd ed. Wiley & Sons, New York, 1958.

Chamot, E.M., Mason, C.W. *Handbook of Chemical Microscopy,* Vol. II, 2nd ed. Wiley & Sons, New York, 1960.

Colchester, A.C.F., Colchester, N.T.H. The origin of bovine spongiform encephalopathy: The human prion disease hypothesis. *Lancet,* **2005,** *336*(9488): 856–861.

Collings, G. H. *Commercial Fertilizers,* 5th ed. McGraw-Hill, New York, 1955.

Creswell, A.W. Flotation technique on teed ingredients and mixed feeds. *Official Proceedings of the American Association of Feed Microscopists,* 1960.

Doherr, Marcus G. 2007. Brief review on the epidemiology of transmissible spongiform encephalopathies (TSE). *Vaccine,* **2007,** *25*(30): 5619–5624.

European Commission (EC). Commission Directive EC/381/94 concerning certain protection measures with regard to bovine spongiform encephalopathy and the feeding of mammalian derived protein. *Official Journal of the European Communities* **1994,** L *172.* pp 23–24.

European Commission (EC). Commission Regulation EC/999/2001 laying down rules for the prevention, control and eradication of certain transmissible spongiform encephalopathies. *Official Journal of the European Communities,* **2001,** L*147.* pp 1–40.

European Commission (EC). Commission Regulation EC/1774/2002 laying down health rules concerning animal by-products not intended for human consumption. *Official Journal of the European Communities,* **2002,** L273. pp 1–95.

European Commission. The TSE roadmap. European Commission, Brussels, 2005, p 24.

European Commission (EC). 2008. Commission Regulation EC/956/2008 amending Annex IV to Regulation (EC) No 999/2001 of the European Parliament and of the Council laying down rules for the prevention, control and eradication of certain transmissible spongiform encephalopathies. *Official Journal of the European Communities* L260. pp 8–11.

European Commission (EC). 2009. Commission Regulation EC/152/2009 laying down the methods of sampling and analysis for the official control of feed. *Official Journal of the European Communities* L54. pp 1–130.

Feigl, F. *Spot Tests in Inorganic Analysis,* 5th ed. Elsevier, New York, 1958.

Feigl, F. *Spot Tests in Organic Analysis,* 6th ed. Elsevier, New York, 1960.

Horwitz, W., ed. Official *Methods of Analysis of the Association of Official Agricultural Chemists (AOAC),* 9th ed. AOAC, Gaithersburg, MD, 1960.

Jacob, K. D., ed. *Fertilizer Technology & Resources in the U.S.* Academic Press, New York, 1953.

Makowski, J.V. Feed microscopy: An undervalued quality assurance technique. *inform,* **1998,** 9(11), 1036–1042.

Manual of Microscopic Analysis of Feeding Stuffs. American Association of Feed Microscopists, 1962.

OIE – World Organization of Animal Health. 2007. Bovine Spongiform Encephalopathy. Accessed August 5, 2010 from: http://www.oie.int/eng/info/en_esbru.htm

OIE – World Organization of Animal Health. 2010. Number of reported cases of bovine spongiform encephalopathy (BSE) in farmed cattle worldwide*(excluding the United Kingdom). Accessed August 5, 2010 from: http://www.oie.int/eng/info/en_esbmonde.htm

Prusiner, S.B. Prions, *Proc Natl Acad Sci,* **1998,** *95*(23):13363–13383.

Sandell, E.B. *Colorimetric Determination of Traces of Metals,* 2nd ed. Interscience Pub, New York, 1950.

Sauchelli, V., ed. *Chemistry & Technology of Fertilizers.* Reinhold, New York, 1960.

Schaeffer, H.F. *Microscopy for Chemists.* Dover, New York, 1953.

Telling, G.C., Scott, M., Mastrianni, J., Gabizon, R., Torchia, M., Cohen, F.E., DeArmond, S.J., Prusiner, S.B. Prion propagation in mice expressing human and chimeric PrP transgenes implicates the interaction of cellular PrP with another protein. *Cell,* **1995,** *83:* 79–90.

USDA, NRCS. 2010. The PLANTS Database (http://plants.usda.gov, 30 July 2010). National Plant Data Center, Baton Rouge, LA 70874-4490 USA.

Veys, P. and Baeten, V. (2010). New approach for the quantification of processed animal proteins in feed using light microscopy. *Food Additives & Contaminants: Part A* **27**(7): 926–934.

Whitehead, A.J. Ensuring food quality and safety and FAO technical assistance. *Food, Nutrition and Agriculture,* **1998,** *21*:10–17.

World Health Organization (WHO). 2002. Fact Sheet N° 113. Bovine Spongiform Encephalopathy. Accessed August 5, 2010 from: http://www.who.int/mediacentre/factsheets/fs113/en/

World Health Organization (WHO). 2002b. Fact Sheet N° 180. Variant Creutzfeld-Jakob Disease. Accessed August 5, 2010 from: http://www.who.int/mediacentre/factsheets/fs180/en/

Wilesmith, J.W., J.B. Ryan, and M.J. Atkinson. Bovine spongiform encephalopathy: epidemiological studies on the origin. *Veterinary Record,* **1991,** *128*(9): 199–203.

Recommended Sources of Information

ARIES – Animal Remains Identification and Evaluation System V.0.1. (2004). RIKILT Institute for food safety, Wageningen, The Netherlands.

Klein, H., Marquard, R. *Feed Microscopy: Atlas for the Microscopic Examination of Feed Containing Vegetable and Animal Products.* Agrimedia GmbH, Bergen, Germany, 2005.

Moore, T.D., Liter E. Dugnolle S., Dugnolle, C.E. *Identification of the Dorsal Guard Hairs of Some Mammals of Wyoming.* Wyoming Game and Fish Department, 1974.

Olsen, A.R., Sidebottom, T.H.E., Knight, S.A., eds. *Fundamentals of Microanalytical Entomology. A Practical Guide to Detecting and Identifying Filth in Foods.* New York, CRC Press, 1996.

Contributors

James Makowski

James Makowski received his Ph.D. from the University of Delaware in Genetics in 1991. He is professor of genetics and microbiology at Messiah College and owner of Winsor Laboratories, a leading forensic microscopy laboratory located in Mechanicsburg, PA. An agricultural microscopist for over 25 years, he has served as president of the American Association of Feed Microscopists (AAFM), currently known as the AOCS Agricultural Microscopy Division, and is a recipient of their coveted Golden Microscope Award for outstanding contributions to the promotion of Agricultural Microscopy. James chairs the Agricultural Microscopy Division's training committee and has held several board and committee positions for AOCS.

Neil Vary

Neil Vary graduated from Carleton University (Ottawa, Canada) with a B.S. (Honors) in Biology and Biotechnology in 2001. Shortly afterwards, he was employed as a biologist within the Feed and Fertilizer Section of the Ottawa Laboratory (Carling) of the Canadian Food Inspection Agency (CFIA). At this time Neil was introduced to the field of feed and fertilizer microscopy. As a biologist, Neil contributed to the development of microscopy and molecular biology methods to provide analytical support for the Canadian ruminant feed ban legislation that was created to help prevent the proliferation of bovine spongiform encephalopathy (BSE; or "mad cow disease"). Neil gained much expertise in the methodologies and techniques that can be employed to detect and identify processed animal proteins in feeds and feed ingredients as well as agricultural microscopy as a whole. He has provided analytical support to numerous food safety and criminal investigations and has provided expert witness testimony within the Canadian judicial system.

Neil has been involved with the Agricultural Microscopy Division (AMD) of the American Oil Chemists' Society (AOCS) since 2001. He was the AMD newsletter editor for two years, vice-chairperson for two years and chairperson for a year. He has also helped develop and instruct several feed microscopy short courses that were offered at several locations throughout North America for the AMD.

Marjorie McCutcheon

Marjorie McCutcheon graduated from West Virginia University (Morgantown, West Virginia) with a B.S. (Honors) in Animal and Veterinary Science in 1981. Marjorie joined the West Virginia Department of Agriculture's current Regulatory and Environmental Affairs Division (READ) in 1982 initially doing feed and fertilizer analysis for consumer protection purposes in what is now known as the READ Agricultural Materials Laboratory. Her strong animal nutrition background coupled with a rich background in plant and soil science made this an excellent career opportunity. This is where her love for microscopy was developed as well. In 1986 she became the WV State Seed Analyst and Feed Microscopist and currently is in charge of all of the seed and feed microscopy programs for the West Virginia Department of Agriculture. These consumer protection based programs include doing germination, identification, purity, and noxious weed analysis on seed samples as well as microscopic identification of feed ingredients and contaminants. Over the years she has done food and fertilizer microscopy as well. She has provided analytical support to criminal investigations in the United States involving seed and animal feed as an expert in her field.

Marjorie is a 29-year veteran microscopist and served ten years as past secretary/treasurer of the American Association of Feed Microscopists and the AOCS Feed Microscopy Division now known as the AOCS Agricultural Microscopy Division. She is past vice-chair and chair of the Agricultural Microscopy Division and is the recipient of their coveted Golden Microscope Award for outstanding contributions to the promotion of feed microscopy.

Pascal Veys

Pascal Veys graduated from the Catholic University of Louvain with a Ph.D. in biology where he was a teaching assistant. Currently, he is head of the microscopy laboratory at the Valorisation of Agricultural Products Department of the Walloon Agricultural Research Centre – CRA-W in Belgium. He works for the European Union Reference Laboratory for Animal Proteins in feedingstuffs – EURL–AP where he is in charge of the coordination of the EURL–AP activities for all National Reference Laboratories of the EU member states. Pascal is responsible for the tutorials and workshops organized in the EU for the detection of processed animal proteins in feed according to the official microscopic method. He is also in charge of the organization of the EURL–AP collaborative studies. His team developed an online micrographs collection. He also participated in the European SAFEED PAP project.

Aside from his membership to the AOCS Agricultural Microscopy Division, he is also a member of the IAG Section Feedingstuff Microscopy and the EMS.

Index

A
Abutilon theophrasti Medik, 131
Activated sludge, 112
Agaliya, 118
Agricultural microscopy
instruments/equipment, 2
principles of, 1–2
reference ingredient collection and, 2
Agricultural Microscopy Division (AOCS), training by, 3
Agrostemma githago, 119
Aleppo milletgrass, 124
Aleurone, 9
Alfalfa products (*Medicago sativa*)
definitions, 33
histological features, 34
macroscopic features, 33–34
stereoscopic features, 34
Alizarin Red
bone and, 79
cartilage and, 85
lacunae/canaliculi and, 81
Allium canadense, 132
Aluminum, 111
Aluminum sulfate, 108
Amaranth family, 127
Amaranthus sp., 127
Ambrosia artemisiifolia, 128
Ambrosia trifida, 129
American pokeweed, 128
Ammoniated superphosphate, 107
Ammonium nitrate, 107, 112
Ammonium nitrate limestone, 107
Ammonium polyphosphate, 107
Ammonium sulfate, 153–154
fertilizer microscopy and, 107, 112
Ammonium sulfate-nitrate, 107
Ammoniums, 110
Amprolium, 162–163
Analytes, microchemical spot tests and, 162–165
Animal liver meal, 54
Animal markers, 96–97
Animal origins. *See* Feed ingredients of animal origin
Anions, 110
Annual ragweed, 128
Apple pomace, 40–41
Arctium minus Bernh, 118
Arsanilic acid, 162–163
Arsonic acid, 162–163
Asclepias syriaca, 125
Ascorbic acid, 164
Association of American Feed Control Officials (AAFCO), 115
Aster family, 118, 122, 128–129
Avena fatua, 132
Avian, bone fragments. *See* Poultry products

B
Bakery product, dried, 39
Balances, 3
Ballmustard, 116
Balloonvine, 116
Barbados nut, 126
Barbane, 118
Barbules, feathers and, 86–87

Barley products (*Hordeum vulgare*)
  definitions, 10
  histological features, 11
  macroscopic features, 10
  stereoscopic features, 10
Barnyardgrass, 116
Basic slag, 109
Bedstraw, 117
Beet pulp, dried, 39
Beggar's button, 118
Benzidine, 110
Bigpod sesbania, 119
Bird's nest, 131
Birefringence
  feathers and, 87–88
  temperature effect on, 81
Birefringence of gills, 57–58
Black bindweed, 131
Blood characteristics
  for detection of animal products in feed, 74–75
  microscopy interpretations of, 95–96
  spot testing of, 165
Blood meal
  definitions, 51
  histological features, 52
  macroscopic features, 51
  stereoscopic features, 51–52
Bone characteristics
  for detection of animal products in feed, 79–84
  microscopy interpretations of, 79–84
  spot testing of, 165
Bone meal
  definitions, 50
  fertilizer microscopy and, 104
  histological features, 50
  macroscopic features, 50
  mammalian, 43–47
  stereoscopic features, 50–51
Borax (sodium borate), 108
Boron, 111
Bottlegrass, 122
Bovine spongiform encephalopathy (BSE), 65–67
Bovines
  hair characteristics of, 91, 94–95
  spot testing of, 165
Bran
  corn, 12–13
  rice, 18–19
  rye, 20
  wheat, 21–22
Brewer's products
  definitions, 34–35
  histological features, 35
  macroscopic features, 35
  stereoscopic features, 35
Broadleaf plantain, 117
Buckhorn plantain, 117
Buckwheat family, 120, 125, 131
Buckwheat feed, 35–36
Bull nettle, 124
Bull thistle, 124
Burdock, 118
Butterprint, 131
Buttonweed, 131

C

Calcium
  definition/features of, 135–137
  spot testing of, 110
Calcium carbonate
  definition/features of, 135–136
  fertilizer microscopy and, 108
Calcium chloride, 135
Calcium iodate, 140–141
Calcium nitrate, 107
Calcium sulfate, 135–137
Cane molasses, 40
Cannabis sativa, 123
Canola, 126
Canola meal (*Brassica* spp.)
  definitions, 23
  histological features, 24
  macroscopic features, 23
  stereoscopic features, 24
Capsella bursa-pastoris, 130
Carbadox, 162–163
Carbonates, 110
Cardiospermum halicacabum, 116
Carelessweed, 127
Carolina nettle, 124
Carrot family, 131–132
Cartilage characteristics
  for detection of animal products in feed, 84–85
  microscopy interpretations of, 96
Cassia obtusifolia, 130
Castor pomace, 105, 112
Castorbean, 118
Catchweed bedstraw, 117

Cations, 110–111
Cats. *See* Felines
Cattail grass, 123
Cattle. *See* Bovines
Cellular/subcellular structure identification, 1
Chaff/dust, 42
Charlock mustard, 126
Chelated iron, 108
Chenopodium album, 125
Chile saltpeter, 107
Chlorides, 164
  spot testing of, 110
Chlorophyll, 34
Chlortetracycline HCl, 162–163
Cirsium discolor, 122
Citrus molasses, 40
Citrus products
  definitions, 36
  histological features, 37
  macroscopic features, 36
  stereoscopic features, 36
Cleavers, 117
Cleaverwort, 117
Climbing buckwheat, 131
Clovers, 34
Cobalt
  definition/features of, 137–138
  spot testing of, 111, 164
Cobalt carbonate, 137–138
Cobalt sulfate, 137
  fertilizer microscopy and, 108
Cocklebur, 118
Cockspur grass, 116
Coconut meal (*Cocos nucifera*), 37
Coffee senna, 119
Coffeebean, 119
Collagen, cartilage and, 84
Colloidal phosphate, 109
Columbus grass, 124
Common burdock, 118
Common cocklebur, 118
Common corncockle, 119
Common lambsquarters, 125
Common milkweed, 125
Common plantain, 117
Common pokeweed, 128
Common ragweed, 128
Common St. John's wort, 129
Compound microscope, 2–3
  and detection of animal products in feed (flotation), 72
  and detection of animal products in feed (sediment), 73–74
  inspection after flotation separation procedure, 4–5
Convolvulus arvensis, 121
Copper
  definition/features of, 138–140
  spot testing of, 111
Copper oxide, 138, 141
  fertilizer microscopy and, 108
Copper sulfate, 138–140
  fertilizer microscopy and, 108, 113
Corn distillers products, 37–38
Corn/maize products (*Zea mays*)
  definitions, 11–12
  histological features, 14
  macroscopic features, 12–13
  stereoscopic features, 13–14
Cornbind, 131
Corncockle, 119
Cotton burr ash (cottonseed hull ash or carbonate of potash), 104–105
Cottonseed meal, 105
Cottonseed products (*Gossypium hirsutum*)
  definitions, 24
  histological features, 25
  macroscopic features, 24–25
  stereoscopic features, 24–25
Cow cockle, 120
Cow soapwort, 120
Cowherb, 120
Crab meal, 61
Creeping jenny, 121
Creutzfeldt-Jakob disease (vCJD), 67
Crotalaria spectabilis, 130
Croton, 120
Croton sp., 120
Curly dock, 120
Cuscuta sp., 121
Cystine, 165

D

Dairy products
  definitions, 52
  histological features, 53
  macroscopic features, 52–53
  stereoscopic features, 53
Darnel ryegrass, 121
Datura stramonium, 124
Daucus carota, 131
Deer/elk, hair characteristics of, 93–95

Defatted germ meal, 21
Defluorinated phosphate, 149
Denticles, fish meal/solubles and, 59–60
Detection of animal products in feed. *See also* Feed ingredients of animal origin
  background, 65–67
  blood characteristics, 74–75
  bone characteristics, 79–84
  cartilage characteristics, 84–85
  examination and, 71–74
  feather characteristics, 85–88
  hair characteristics, 88–95
  microscopy interpretations of blood, 95–96
  microscopy interpretations of bones, 96
  microscopy interpretations of feathers, 96
  microscopy interpretations of hair, 96–97
  microscopy interpretations of milk products, 96
  microscopy interpretations of multiple animal markers, 97
  microscopy interpretations of muscle tissue, 96
  milk product characteristics, 75–77
  muscle tissue characteristics, 77–78
  North America/Europe and, 65
  sample preparation for, 68–71
  and value of feed microscopy, 67–68
Diammonium phosphate, 107
Dicalcium/monocalcium phosphate, 149–150
Differential interferential contrast (DIC), muscle tissue characteristics and, 78
Dimetridazole, 162–163
Diseases, 65–67
Distillers products
  definitions, 37–38
  histological features, 38–39
  macroscopic features, 38
  stereoscopic features, 38
DMAB, spot testing of, 110
DNA-based methods of microscopy, benefits of, 67–68
Dodder, 121
Dodder family, 121
Dolomite (dolomitic limestone), 108–109
Double/triple superphosphate, 107
Doveweed, 120
Dried bakery product, 39
Dried beet pulp, 39
Dullseed cornblind, 131

E

Echinochloa crusgalli, 116
Education/training, AOCS and, 3
Egg shell meal, 47. *See also* Poultry products
Elastic cartilage, 84
Elemental sulfur, 153–154
Emjeo, 108
Emtryl, 162–163
Endosperm, 9
English plantain, 117
Epsom Salts, 108
Equipment. *See* Instruments/equipment
Ethylenediamine dihydroiodide (EDDI), 140–142
European bindweed, 121
Evaporating dishes, 3
Extraction
  cottonseed products and, 24
  safflower products and, 28–29
  soybean products and, 30
  sunflower products and, 32

F

Fanweed, 122
Feather characteristics
  for detection of animal products in feed, 85–88
  microscopy interpretations of, 96
Feathermeal tankage, 104
Feathers, 165
Feed ingredients of animal origin
  animal liver meal, 54
  blood meal, 51–52
  bone meal, 50–51
  dairy products, 52–53
  mammalian meat meal/meat and bone meal, 43–47
  meat meal tankage/meat and bone meal tankage, 53–54
  poultry products, 47–50
Feed ingredients of marine origin
  crab meal, 61
  fish meal/solubles, 55–61
  invertebrate meals, 63–64
  shrimp/krill meals, 61–63
Feed ingredients of plant origin
  alfalfa products (*Medicago sativa*), 33–34
  barley products (*Hordeum vulgare*), 10–11
  Brewer's products, 34–35
  buckwheat feed, 35–36
  canola meal (*Brassica* spp.), 23–24

cereal grain structure, 9–10
citrus products, 36–37
coconut meal (*Cocos nucifera*), 37
corn/maize products (*Zea mays*), 11–14
cottonseed products (*Gossypium hirsutum*), 24–25
distillers products, 37–39
dried bakery product, 39
dried beet pulp, 39
flaxseed/linseed products (*Linum usitatissimum L.*), 26–27
grain sorghum products (*Sorghum bicolor M.*), 14–15
kelp meal, 39
molasses products, 40
oat products (*Avena sativa*), 16–18
oilseeds, 23
peanut products (*Arachis hypogaea L.*), 27–28
pomaces, 40–41
quantitative estimation by flotation/separation, 5–7
rice products (*Oryza sativa*), 18–19
rye products (*Secale cereale L.*), 20–21
safflower products (*Carthamus tinctorius*), 28–29
screenings, 41–42
soybean products (*Glycine max*), 30–32
sunflower products (*Helianthus annus L.*), 32–33
triticale, 42
wheat products (*Triticum* sp.), 21–23
Felines, 94–95
Ferrous carbonate, 142–144
Ferrous sulfate, 142–144
fertilizer microscopy and, 108
Fertilizer microscopy
activated sludge and, 112
ammonium nitrate and, 112
ammonium sulfate and, 112
bone meal and, 104
castor pomace and, 105, 112
copper sulfate and, 113
cotton burr ash (cottonseed hull ash or carbonate of potash), 104–105
cottonseed meal and, 105
economics/quality and, 99
excess chlorine and, 100
feathermeal tankage and, 104
guano and, 106
identification using smell, 102–103
inorganic materials and, 107–109
iron sulfate and, 113–114
leathermeal tankage and, 104
manures and, 106
packinghouse tankage and, 103–104
peanut hull meal and, 105
peat/muck and, 106
Peruvian guano and, 112
potassium chloride and, 113
purity of nitrogen source for, 99–100
rice hulls and, 105
sewage sludge and, 103
soybean meal and, 105–106
superphosphate and, 112–113
tankage and, 103
techniques/equipment for, 101–102
tobacco stems and, 104
tung meal and, 112
tung pomace and, 105
urea and, 106, 113
urea-formaldehyde and, 106–107
value to manufacturers, 100–101
Fibrous cartilage, 84
Field bindweed, 121
Field parsnip, 132
Field pennycress, 121
Field thistle, 121
Fines, 33
Fish
bone fragments, 80–84
meal/solubles, 55–61
Fiveangled dodder, 121
Flaxgrass, 132
Flaxseed/linseed products (*Linum usitatissimum L.*)
definitions, 26
histological features, 26–27
macroscopic features, 26
stereoscopic features, 26
Flotation
and detection of animal products in feed with compound microscope, 72
and detection of animal products in feed with stereoscope, 71–72
quantitative estimation of ingredients by, 5–7
sample separation and, 3
Flotation and sample separation, 69–71
alfalfa products and, 34
Flotation concentrates, 109
Frenchweed, 122
Furazolidone, 162–163

G
Galeopsis tetrahit, 123
Galium aparine, 117
Gelatin, 165
Gentian violet, 162–163
Giant ragweed, 129
Gills, 55–56
Gluten
  corn, 12–14
  sorghum, 14–15
Goosefoot family, 125
Gossypol, 25
Grain screenings, 41–42
Grain sorghum products (*Sorghum bicolor M.*)
  definitions, 14
  histological features, 15
  macroscopic features, 15
  stereoscopic features, 15
Grape pomace, 40–41
Grass, 123
Grass family, 116, 121–124, 127, 129, 132
Great ragweed, 129
Green bristlegrass, 122
Green foxtail, 121
Green millet, 122
Grits, corn, 12
Groats, 16
Guano, 106, 112

H
Hair characteristics
  for detection of animal products in feed, 88–95
  microscopy interpretations of, 96–97
Hairs
  bone meal and, 45
  general histological features by species, 94–95
  spot testing of, 165
Hashish, 123
Hay, 27–28
Hemicellulose extract, 40
Hemp, 122
Hemp family, 123
Hemp sesbania, 119
Hempnettle, 122
Herbe de cuba, 124
Hilum, 31
Hogs. *See* Porcines
Hogwort, 120
Hominy feed, 12
Hooves, 165
Horns, 165
Horse products, 67–68
  spot testing of, 165
Horsenettle, 124
Horseweed, 129
Hour glass cells of soybean, 31–32
Humans, hair characteristics of, 94–95
Hyaline cartilage, 84
Hypericum perforatum, 129

I
Indian mallow, 131
Inkberry, 128
Inorganic material identification in fertilizer, 107–109
  spot testing of, 109–111
Instruments/equipment
  basic, 3
  compound microscope, 2–3
  stereoscope, 2
Invertebrate meals, 63–64
Iodate, 164
Iodide, 164
Iodine, 140–142
Ipomoea sp., 126
Iron
  definition/features of, 142–144
  spot testing of, 111, 164
Iron, chelated, 108
Iron oxide, 143
Iron sulfate, 108, 113–114

J
Jamestown weed, 124
Jatropha sp., 126
Java bean, 130
Jimsonweed, 124
Johnsongrass, 124

K
Kaber mustard, 126
Kelp meal, 39
Kibbled corn, 12
Klamathweed, 129
Kneegrass, 127
Krill meal, 61–63

L
Lactoserum powder, 75–77
Lacunae

bone meal and, 45
fish bone and, 80
fish meal/solubles and, 55–57
Lady's Thumb, 125
Ladysthumb, 125
Lama palagi, 118
Lambsquarters, 125
Lanceleaf plantain, 117
Leathermeal tankage, 104
Leatherstem, 126
Legumes, 23
Lenses of fish eyes, 60–61
Lesser burdock, 118
Lily family, 132
Limestone (calcium carbonate)
fertilizer microscopy and, 108
spot testing of, 165
Linseed. *See* Flaxseed/linseed products (*Linum usitatissimum L.*)
Liver meal, 54
Lolium temulentum, 121
Low ragweed, 128

M

Macroscopic identification, 1
Mad apple, 124
Mad cow disease. *See* Bovine spongiform encephalopathy (BSE)
Madder family, 117
Magnesium
definition/features of, 144–146
spot testing of, 110, 164
Magnesium ammonium phosphate, 107
Magnesium oxide, 144–146
Magnesium sulfate anhydrous, 145–146
Magnesium sulfate (Emjeo/Epsom Salts), 108
Maize. *See* Corn/maize products (*Zea mays*)
Mallow family, 128, 131
Malt sprouts/cleanings, 35
Mammalian meat and bone meal
definition, 43
histological features, 44–47
macroscopic features, 43–44
stereoscopic features, 44
Mammalian meat meal, 43
Manganese
definition/features of, 146–147
spot testing of, 110, 164
Manganese oxide, 146–148
fertilizer microscopy and, 108
Manganese sulfate, 147–148
fertilizer microscopy and, 108
Mangosteen family, 129
Manures, 106
Marijuana, 123
Markers, 68
Mary jane, 123
Maskerekur, 118
Meadow garlic, 132
Meal, corn, 12
Meat and bone meal. *See* Mammalian meat and bone meal
Meat meal tankage/meat and bone meal tankage, 53–54
Melengestrol acetate, 162–163
Menadione, 164
Menadione sodium bisulphate, 164
Methods of microchemical spot tests. *See* Microchemical spot tests
Mice, hair characteristics of, 93–95
Microchemical spot tests
for animal products, 165
for drugs (color reaction/detection limits in chop feed), 162–163
general procedures, 161, 165
Method I, 162, 164–166
Method II, 162, 164, 166
Method III, 162, 164, 166
Method IV, 162, 164, 166
Method IX, 164, 167
Method V, 162, 164, 166
Method VI, 162, 164, 166
Method VII, 164–165, 167
Method VIII, 164, 167
Method X, 164, 167
Method XI, 164, 167–168
Method XII, 168
Method XIII, 162, 168
Method XIV, 168
Method XIX, 162, 169
Method XV, 165, 169
Method XVI, 165, 169
Method XVII, 164, 169
Method XVIII, 164, 169
Method XX, 165, 169–170
Method XXI, 165, 170
Method XXII, 165, 170
for minerals (color reaction), 164
MSDS and handling/storage, 161
for vitamins (color reaction), 164
Micrometer, ocular, 3

Microscopes
compound microscope, 2–3, 4–5, 72
interpretations/limitations of, 95–97
Microscopy interpretations of blood, 95–96
Microscopy interpretations of bones, 96
Microscopy interpretations of feathers, 96
Microscopy interpretations of hair, 96–97
Microscopy interpretations of milk products, 96
Microscopy interpretations of multiple animal markers, 97
Microscopy interpretations of muscle tissue, 96
Middlings
rye, 20
wheat, 21
Milk. *See* Dairy products
Milk product characteristics
for detection of animal products in feed, 75–77
microscopy interpretations of, 96
Milkweed, 125
Milkweed family, 125
Minerals
calcium, 135–137
cobalt, 137–138
copper, 138–140
iodine, 140–142
iron, 142–144
magnesium, 144–146
manganese, 146–147
phosphorus, 147–150
potassium, 150–152
selenium, 152–153
sulfur, 153–154
zinc, 155–159
Mint family, 123
Mixed screenings, 42
Molasses distillers products, 37–38
Molasses products, 40
Mollusks, 63
Molybdate solution, 109
Molybdenum, 111
Monoammonium phosphate, 149–151
Monensin Na, 162–163
Monosodium phosphate, 149–150
Moonflower, 124
Morantel tartrate, 162–163
Morning-glory, 126
Morning-glory family, 121, 126
Mortar and pestle, 3
Multiple animal markers, 96–97
Muriate of potash, 108
Muscle tissue characteristics
for detection of animal products in feed, 77–78
microscopy interpretations of, 96
spot testing of, 165
Mussel shell powder, 63–64
Mustard, 126
Mustard family, 116, 122, 126, 130, 133

N

Nacre, 63–64
Narrowleaf dock, 120
Narrowleaf plantain, 117
Nassella trichotoma, 129
Nassella tussock, 129
Near infrared microscopy (NIRM), 67
Needles, 3
Neslia paniculata, 116
Nessler's reagent, 109–110
Nettlespurge, 126
Nifursol, 162–163
Nightflowering catchfly, 127
Nightshade family, 124
Nitrate of soda, 107
Nitrates, 110, 164
Nitrofurazone, 162–163
Nucellar layer, 9
Nucellus, 9

O

Oat products (*Avena sativa*)
definitions, 16
histological features, 17–18
macroscopic features, 16
stereoscopic features, 16–17
Oatgrass, 132
Ocular micrometer, 3
Oilseeds, 23
definition, 23
Oleandomycin, 162–163
Organic material identification in fertilizer
bone meal and, 104
castor pomace and, 105
cotton burr ash (cottonseed hull ash or carbonate of potash), 104–105
cottonseed meal and, 105
feathermeal tankage and, 104

guano and, 106
leathermeal tankage and, 104
manures and, 106
packinghouse tankage and, 103–104
peanut hull meal and, 105
peat/muck and, 106
rice hulls and, 105
sewage sludge and, 103
soybean meal and, 105–106
tankage and, 103
tobacco stems and, 104
tung pomace and, 105
urea and, 106
urea-formaldehyde and, 106–107
Otolith fragments, 57–58
Ovines, 93–95
Oxytetracycline HCl, 162–163
Oyster powder, 63–64

P

Packinghouse tankage, 103–104
Panicgrass, 127
Panicum, 127
Panicum sp., 127
Parsnip, 132
Pastinaca sativa, 132
Pea family, 119, 130
Peanut hull meal, 105
Peanut products (*Arachis hypogaea L.*)
definitions, 27
histological features, 28–29
macroscopic features, 27
stereoscopic features, 27–28
Pearl barley, 10
Peat/muck, 106
Pebble phosphate, 109
Penicillin, 162–163
Pennycress, 122
Perennial morning-glory, 121
Pericarp, 10
Peruvian guano, 112
Phenylhydrazine, 110
Phosphate rock, 109
Phosphate spiral, 109
Phosphate tailings, 109
Phosphates, 110, 164
Phosphorus, 147–150
Phytolacca Americana, 128
Pig products, 67–68
Pigeonberry, 128
Pigeongrass, 122–123
Pigweed, 127
Pink family, 119–120, 127
Piperazine, 162–163
Plant origins. *See* Feed ingredients of plant origin
Plantago lanceolata, 117
Plantago major, 117
Plantain family, 117
Pokeberry, 128
Pokeweed, 128
Pokeweed family, 128
Polishings, rice, 18–19
Polygonum convolvulus, 131
Polygonum persicaria, 125
Pomaces, 40–41
fertilizer microscopy and, 105–106
Porcines, 91–92, 94–95
Potassium
definition/features of, 150–152
spot testing of, 110
Potassium chloride, 123, 150–152
fertilizer microscopy and, 108, 113
Potassium iodide, 140–142
Potassium/magnesium sulfate, 151–152
Potassium nitrate, 108
Potassium polyphosphate, 107
Potassium sulfate, 151
fertilizer microscopy and, 108
Potato family, 124
Poultry products
bone fragments, 82–83
definitions, 47
histological features, 49–50
macroscopic features, 47
stereoscopic features, 47–48
Press liquor, 55
Prickly fanpetals, 128
Prickly sida, 128
Processed animal product (PAP), 68
Purghera, 126
Purging nut, 126
Purple cockle, 119

Q

Quantitative estimation of ingredients
by flotation/separation, 5–7
meat meal/poultry meal separations, 6
typical complete separations, 6
typical finished feed formula, 7

Quantitative work, 5
Queen Anne's lace, 131

R

Ragweed, 128
Rapeseed, 126
Raphanus raphanistrum, 133
Rations
  meat meal/poultry meal separations, 6
  typical complete separations, 6
  typical finished feed formula, 7
Rattlebox, 130
Reagents, 162–165
Ribgrass, 117
Riboflavin, 164
Ribwort, 117
Rice hulls, 105
Rice products (*Oryza sativa*)
  definitions, 18
  histological features, 19
  macroscopic features, 18–19
  stereoscopic features, 19
Ricinus communis, 118
Rippleseed plantain, 117
Roadside thistle, 122
Robendine HCl, 162–163
Rodents, 93–95
Roman wormwood, 128
Rough cocklebur, 118
Rumex crispus, 120
Rye distillers products, 38–39
Rye products (*Secale cereale L.*)
  definitions, 20
  histological features, 20–21
  macroscopic features, 20
  stereoscopic features, 20

S

Safety, separations with chloroform/petroleum ether and, 6
Safflower products (*Carthamus tinctorius*)
  definitions, 28–29
  histological features, 29
  macroscopic features, 29
  stereoscopic features, 29
Saint Johnswort, 129
Salinomycin, 162–163
Sample preparation, 3
  for detection of processed animal products, 68–71
Sample separation
  by flotation, 3
  quantitative estimation of ingredients by, 5–7
  by sieving, 4
Saponaria vaccaria, 120
Scales of fish meal, 55–56, 58
Scarthgrass, 117
Screenings, 41–42
Seaweed. *See* Kelp meal
Sediment, and detection of animal products in feed, 72–74
Selenium, 152–153
Selenium selenite, 152–153
Senna obtusifolia, 130
Senna occidentalis, 119
Septicweed, 119
Serrated tussock, 129
Sesbania herbacea, 119
Setaria pumila, 123
Setaria viridis, 122
Sewage sludge, 103
Sheep. *See* Ovines
Shepherd's purse, 130
Shorts, wheat, 21
Showy crotalaria, 130
Showy rattlebox, 130
Shrimp/krill meals, 61–63
Sicklepod, 130
Sicklepod senna, 130
Sida spinosa, 128
Sieves
  as basic equipment, 3
  sample separation by, 4
Silene noctiflora, 127
Sinapis arvensis, 126
Slag, 109
Sludge, 112
Smallflowered morning-glory, 121
Smartweed, 125
Smooth millet, 123
Soapberry family, 116
Sodium, 110
Sodium borate, 108
Sodium molybdate, 108
Sodium nitrate, 107
Sodium sulfur, 153–154
Sodium tripolyphosphate, 149–150
Solanum carolinense, 124
Sorgho de Alep, 124

Sorghum distillers products, 38
Sorghum halepense, 124
Sour dock, 120
Soybean meal, 105–106
Soybean products (*Glycine max*)
definitions, 30
histological features, 31–32
macroscopic features, 30–31
stereoscopic features, 30–31
Spatulas, 3
Spot tests. *See* Microchemical spot tests; *specific sample, i.e. Amprolium*
Sprouting crabgrass, 127
Spurge family, 118, 120, 126
Squid meal, 63–64
St. John's wort, 129
Stains, Alizarin Red, 79, 81, 85
Starch molasses, 40
Starchy endosperm, 9
Stereoscope. *See also specific product, e.g. Bone characteristics*
compound microscope, 2
and detection of animal products in feed (flotation), 71–72
and detection of animal products in feed (sediment), 72–73
inspection after flotation separation procedure, 4
Stickwater, 55
Sticky cockle, 127
Stinkweed, 122
Stinkwort, 124
Stipa trichotoma, 129
Sulfamethazine, 162–163
Sulfaquinoxaline, 162–163
Sulfate of potash, 108
Sulfate of potash-magnesia (Sul-Po-Mag), 108
Sulfates, 107–108, 112–114, 138–140, 142–148, 151–159
Sulfur
definition/features of, 153–154
fertilizer microscopy and, 108
Sulphates, 110, 164. *See also* Sulfates
Sunflower products (*Helianthus annus L.*)
definitions, 32
histological features, 33
macroscopic features, 32
stereoscopic features, 33
Superphosphate, 107, 112–113
Syringe, 3

T

Tankage
fertilizer microscopy and, 103
meat meal/meat and bone meal, 53–54
Teeth, fish meal/solubles and, 59–60
Temperature
and loss of birefringence, 81
and stereoscopic identification of meat and bone meals, 77
Terrestrial bone, 79
Testa, 9–10
cottonseed products and, 26–27
peanut products and, 28
Tetramethylbenzidine (TMB), 74
Thiamine, 164
Thistle, 122
Thlaspi arvense, 122
Thorn apple, 124
Tobacco stems, 104
Tomato pomace, 40–41
Toto ni vavalagi, 118
Transmissible spongiform encephalopathies (TSEs), 65–66
Trichomes
alfalfa products and, 34
oat, 16–17
Triticale, 42
Tung meal, 112
Tung pomace, 105

U

Uluchula skoki, 118
Uranyl zinc acetate, 110
Urea, 106, 113
Urea-formaldehyde, 106–107

V

Velvetleaf, 131
Vente conmigo, 120
Vermiculite, 109
Virginiamycin, 162–163
Vitamin A, 164

W

Watergrass, 116
Weed seeds
ballmustard, 116
balloonvine, 116

barnyardgrass, 116
bedstraw, 117
broadleaf plantain, 117
buckhorn plantain, 117
burdock, 118
castorbean, 118
cocklebur, 118
coffee senna, 119
coffeebean, 119
corncockle, 119
cow cockle, 120
croton, 120
curly dock, 120
darnel ryegrass, 121
dodder, 121
field bindweed, 121
field pennycress, 121
field thistle, 121
giant ragweed, 129
green foxtail, 121
hemp, 122
hempnettle, 122
horsenettle, 124
importance of identification of, 115
jimsonweed, 124
johnsongrass, 124
ladysthumb, 125
lambsquarters, 125
milkweed, 125
morning-glory, 126
mustard, 126
nettlespurge, 126
nightflowering catchfly, 127
panicum, 127
pigweed, 127
pokeweed, 128
prickly sida, 128
ragweed, 128
Saint Johnswort, 129
serrated tussock, 129
shepherd's purse, 130
showy crotalaria, 130
sicklepod, 130
velvetleaf, 131
wild buckwheat, 131
wild carrot, 131
wild garlic, 132
wild oat, 132
wild parsnip, 132
wild radish, 133
yellow foxtail, 122
Wheat distillers products, 38–39
Wheat, grain composition of, 9–10
Wheat oats, 132
Wheat products (*Triticum* spp.)
definitions, 21
histological features, 22–23
macroscopic features, 21
stereoscopic features, 21–22
Wheat red dog, 21
Whey products, 76
White goosefoot, 125
White hedge, 117
Wild buckwheat, 131
Wild burdock, 118
Wild carrot, 131
Wild garlic, 132
Wild mustard, 126
Wild oat, 132
Wild onion, 132
Wild parsnip, 132
Wild radish, 133
Wild rhubarb, 118
Wood molasses, 40
Woolly croton, 120

X
Xanthium strumarium, 118

Y
Yellow bristlegrass, 123
Yellow dock, 120
Yellow foxtail, 122
Yellow weed, 116

Z
Zacate johnson, 124
Zinc
definition/features of, 155–159
spot testing of, 110–111, 164
Zinc oxide, 155–159
Zinc sulfate, 155–159
fertilizer microscopy and, 108
Zoalene, 162–163